JAPAN'S IMPERIAL FOREST, *GORYŌRIN*, 1889–1946

JAPAN'S IMPERIAL FOREST, *GORYŌRIN*, 1889–1946

WITH A SUPPORTING STUDY OF THE *KAN/MIN* DIVISION OF WOODLAND IN EARLY MEIJI JAPAN, 1871–76

by

CONRAD TOTMAN

GLOBAL
ORIENTAL

JAPAN'S IMPERIAL FOREST, *GORYŌRIN*, 1889–1995

First published in 2007 by
GLOBAL ORIENTAL LTD
PO Box 219
Folkestone
Kent CT20 2WP
UK

www.globaloriental.co.uk

ISBN 978-1-905246-30-4

British Library Cataloguing in Publication Data
A CIP catalogue entry for this book is available from the British Library

Set in 9/10.5pt Stone Serif by Servis Filmsetting Ltd, Manchester
Printed and bound in England by Cromwell Press Ltd, Trowbridge, Wilts

FOR

HIDEO KANEKO

Colleague and Friend

The reading for this study on Goryōrin was immensely expedited by Hideo Kaneko's long dedication to developing Yale University's collection of scholarly materials on East Asia. Hideo, who retired as Yale's Curator of the East Asian Collection in August 2000, after thirty-four years of service in the library, was a giant in his profession. During his tenure as Curator, the East Asian Collection tripled its holdings. That growth assured Yale a collection that entered the twenty-first century up to date, rich in content, and sound in organization.

It is in great part thanks to Hideo's labors, and those of other similarly devoted librarians, that scholars have been able to broaden and deepen the English-language corpus of writings on Japan, helping readers of that language to press on in the great enterprise of appreciating the wondrous diversity and complexity of our communal human experience on this little planet.

Contents

Illustrations

Appendices

A Chronology Relating to the History of *Goryōrin*, 1868–1947*

NOTE: Scholarly sources do not always agree on dating. Particularly helpful for clarifying issues of dating in Japan's forest history is the wonderfully detailed work published in 1997 by Nihon ringyō chōsakai, *(Sōgō nenpyō) Nihon no mori to ki to hito no rekishi* [The History of Japan's forests, trees, and people: a chronology]. It covers the centuries from prehistory to the present, with pp. 370–565 treating the decades since 1868.

Date	**Subject**
1868	*Goryōchi* is abolished; Court is funded through government treasury.
1871–73	The Iwakura Mission visits Europe.
March 25, 1872	Palaces are defined as Imperial Property (*kōgūchi*) within the category of Government Property (*kan'yūchi*).
July 1872	A law on title deeds establishes the categories of *kanrin* and *kōūchi*.
Nov. 7, 1874	The government (Dajōkan) cancels earlier decrees on forest organization and issues new regulations for categorizing all land as either government (*kan'yū*) or private (*min'yū*) land. Imperial property is labeled Category 2 of *kan'yūchi*, except for palaces, which are placed in Category 1.
Fall 1875	Matsuno Hazama (Kan) returns to Japan after studying forestry in Germany.
1876	Kido Kōin urges assigning forests to the Imperial Household. From 1879 onward, numerous influential others advocate various ways to equip the Household with woodland and other properties.

Spring 1878	Matsuno et al. establish a forestry research station at Nishigahara in northern Tokyo, under the Forestry Division (Sanrinka) of the Interior Ministry (Naimushō). On December 1, 1882 a Tokyo School of Forestry (Sanrin Gakkō) is established at Nishigahara and is incorporated into Tokyo Imperial University.
1879	The Forest Agency issues a mensuration manual (*Shakujime shidashi benran*); the Property Office uses it during the 1890s.
1880	*Goryōchi* acreage totals 634 *chōbu*.
March 1881	Ōkuma Shigenobu calls for an elected legislature to convene in 1883.
Oct. 12. 1881	An Imperial decree calls for an elected Diet in 1890.
1881–85	Matsukata deflation; peasant rebellions during 1882–84.
Feb. 1882	Itō Hirobumi urges tripartite division of the realm, then departs for Europe with Nakamura Yaroku et al.; returns in August 1883.
March 1884	Itō heads the Imperial Household Ministry (Kunaishō) and the temporary government reform committee.
1884–85	Hokkaido pasturage is assigned to *goryōchi*; with more added in January, 1889.
1885	*Goryōchi* acreage totals 22,000 *chōbu*.
Nov. 28, 1885	At Itō's request, the government begins to restrict sales of larger parcels of government woodland, which are to be used for Goryōrin.
Dec. 22, 1885	Itō takes the added title of Prime Minister (Naikaku Sōri Daijin).
Dec. 23, 1885	The Property Office (Goryōkyoku) is established.
Feb. 1886	The Property Office is expanded; and again in March and September.
Oct. 1887	The Ministry of Agriculture and Commerce (Nōshōmushō) begins focusing its woodland survey work on areas to be assigned to the Property Office.
May 1888	A joint Nōshōmu-Naimu-Kunaishō committee is formed to deal with the reassignment of government woodland (*kokuyūrin*) as *goryōchi*. During 1888–90 it, and a successor committee, identify about 1.6 million *chōbu* to be assigned.
Feb. 11, 1889	The Meiji Constitution and Imperial House Law are issued.

1889	Gunma Prefecture issues regulations for the managing of *goryōchi* there.
1889	The Imperial Household receives Sado and Ikuno mines; disposes of them a few years later.
1889	The Property Office staff numbers thirty-three.
June 1890	The Property Office establishes a regional headquarters in Sapporo. In July the Household asks for more Hokkaido land; the transfer is ordered on August 23, and on September 12 an estimated 2 million *chōbu* are transferred, giving the Household a nominal 3.6 million *chōbu* total acreage.
Sep. 1890	The governor of Yamanashi issues eighteen-clause rules on *goryōchi* leases, in line with instructions from the Property Office.
Nov. 25, 1890	The first Imperial Diet session convenes.
Nov. 28, 1890	The Property Office designates as hereditary (*seden*) *goryō* woodland acreage totalling 1,010,000 *chōbu* in fifteen locations.
Nov. 24, 1892	The Property Office announces its intent to survey its lands, planning to finish the job by 1896. It distinguishes woodland from all other types of *goryōchi*. This Goryōrin shows a profit from this year onward.
April 8, 1893	The governor of Hokkaido seeks return of Goryōrin on the island.
August 1, 1894	Sino-Japanese War erupts.
Oct. 19, 1894	Imperial Household returns 1,370,000 *chōbu* of Hokkaido Goryōrin.
1893–94	The Household Ministry issues decrees to settle Goryōrin boundaries by 1907; surveying is pursued; current uses are to be identified by 1907, Naichi Goryōrin boundaries by 1911, and those of Hokkaido by 1914.
April 27, 1896	A new Civil Law Code is issued, to become effective July 16, 1898. Clause No. 206 defines both *shoyūken* (ownership rights) and *iriaiken* (commonland rights) in woodland.
Sep. 1897	The Property Office establishes five branch headquarters to oversee Goryōrin operations, replacing prefectural officials.
1898–99	The Property Office drafts the first management plan for Goryōrin, with revised rules on mensuration procedure.

Jan. 31, 1898	The Household Ministry issues "Special sales regulations" for *goryōchi*, to allow sale of unwanted parcels. The first set of sales begins this year; three sets are completed by 1936.
May 18, 1898	On the basis of the 1896 law, the Supreme Court (Daishin'in) recognizes *iriaiken* in government land (*kan'yūchi*), and, henceforth, Nōshōmushō hears requests for disposal of such areas.
1899ff.	The Household Ministry promotes the opening of Teshio Plain in Hokkaido.
April 12, 1899	Regulations (*hōritsu* No. 99) on returning government land to private parties are announced, with petitions for return to be due by June 30, 1900.
May 1900	The Household Ministry announces its rules for return of *goryōchi* and timber stands; petitions must be received by June 30, 1901.
August 1900	The eleven-hamlet *iriai* group near Mt. Fuji petitions to regain nearby *goryōrin*, but it is denied. So, in November 1901, the group submits a proposal for use of the land.
April 1901	The Household Ministry says it will allow leasing and harvesting, but not sale, on hereditary (*seden*) Goryōrin.
1901	A forest railway is built in Goryōrin of Adera valley, Kiso.
1901	The Yamanashi prefectural legislature laments forest damage and drafts remedial legislation, issuing new forest regulations in August 1902.
1904–05	Russo-Japanese War.
1905	The Property Office sets up a sawmill in Adera valley.
Feb. 1906	The Property Office establishes its Construction Agency (Dobokuka) for forest projects.
1906	The Property Office establishes a regional headquarters at Kōfu to manage its Yamanashi *goryō*; in June it prepares plans to sell Yamanashi *goryō* parcels with *iriai* rights; in November it announces the sale.
Nov. 1, 1907	The Property Office (Goryōkyoku) is renamed Supervisory Office of the Imperial Forest (Teishitsu Rin'ya Kanrikyoku).
1908–10	The Supervisory Office drafts its second management plan, completing it by May 18, 1910. It begins its second sale of *goryōchi*.

1909	An elaborate sawmill at Nojiri replaces that in Adera.
1910	The Supervisory Office numbers sixty-one officials and fifty-three assistants.
March 11, 1911	The Household abandons Yamanashi *goryō*, following the typhoons of 1907 and 1910 and other difficulties.
1911	The Nojiri sawmill burns.
1913	The Supervisory Office revises its mensuration procedures.
1918	Regulations for the third sale of *goryōchi* are developed; in 1919 the sale is announced.
June 1920	The Household Ministry adopts a payment-in-lieu-of-taxes policy.
1920–21	The Sapporo regional headquarters develops an operating plan for its Sanru area.
Sep. 1, 1923	The Kantō earthquake devastates Tokyo-Yokohama vicinity.
April 9, 1924	The Supervisory Office of the Imperial Forest (Teishitsu Rin'ya Kanrikyoku) shortens its name to Bureau of the Imperial Forest (Teishitsu Rin'yakyoku).
1925	The third *goryō* sale is expanded, in a new effort to cull unwanted parcels.
1926	The floating and rafting of logs on the Kiso River ends.
1927–34	The Bureau of the Imperial Forest drafts its Hundred-Year Plan.
1929	Caterpillar tractors are introduced to the Sanru operations area.
1931	The Household Ministry sells its Nishikori landing.
Sep. 21, 1936	The Household Ministry's Directive No. 6 starts implementing the Hundred-Year plan.
1937	Only 208,478 *chōbu* of *seden goryō* remain (of 1,010,000 claimed in 1890).
1937	A major war with China erupts; expands to involve others in 1941; grinds on until late 1945.
1938	Gasoline rationing starts.
April 1, 1943	The Bureau orders "special harvesting" for the war effort.
April 23, 1943	The Hundred-Year plan is suspended "for the duration."

1946	The Bureau staff totals 1,829; Goryōrin acreage measures 1,349,505 *chōbu*; its stumpage is valued at 2.6 billion *yen*.
Nov. 1946	The consolidation of "public" forest administration commences.
March 31, 1947	Administration of public forests is assigned to the Ministry of Agriculture and Forestry (Nōrinshō). On March 31 it takes over Naichi Goryōrin; on May 1, Hokkaido Goryōrin.
1947	The chain-saw is introduced to Japan and used henceforth.

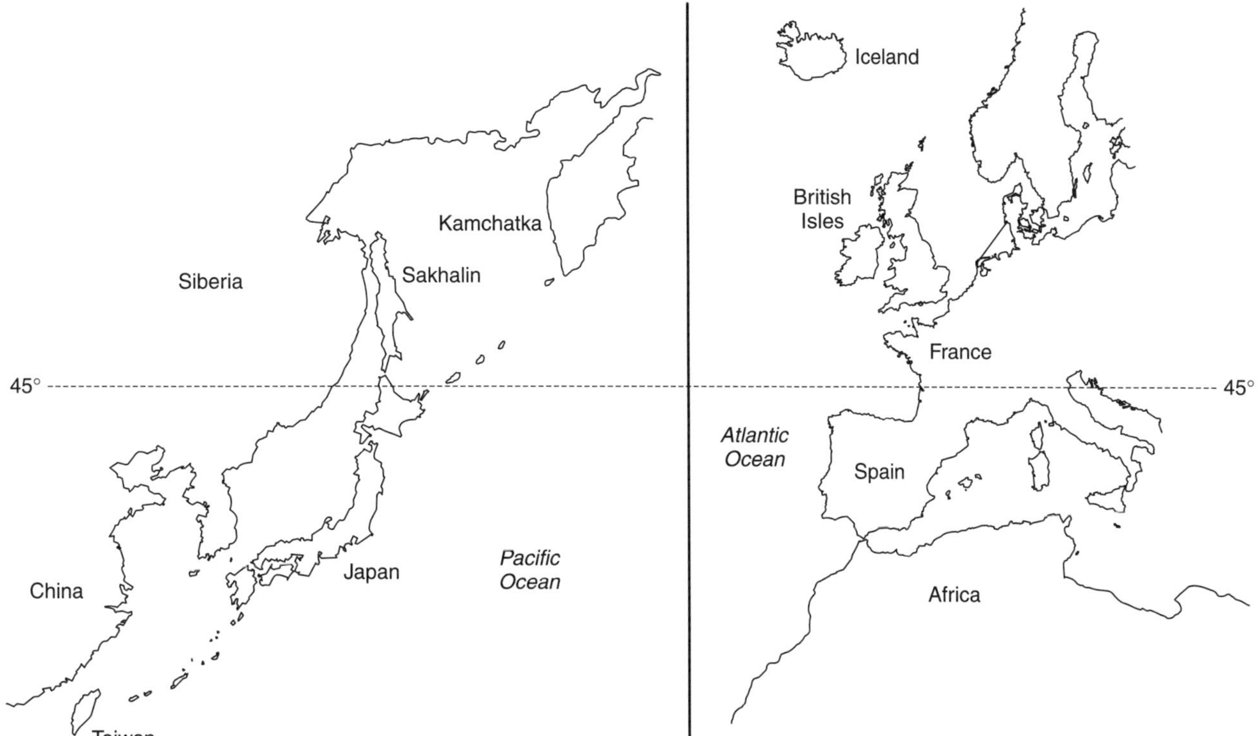

Map 1. Japan and Britain

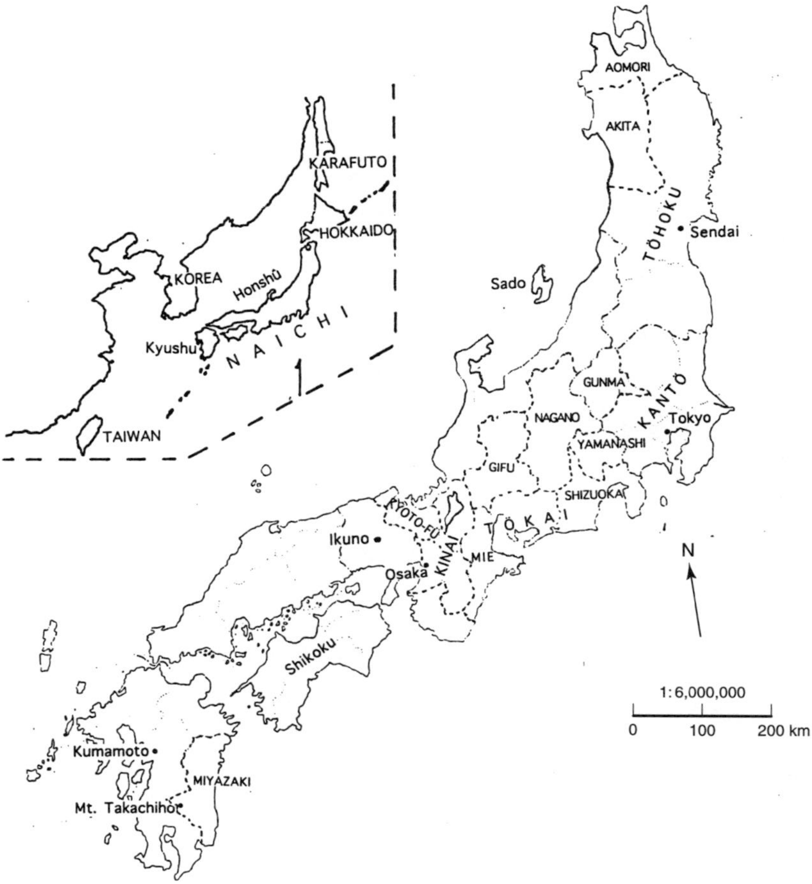

Map 2. Northeast Asia and Naichi

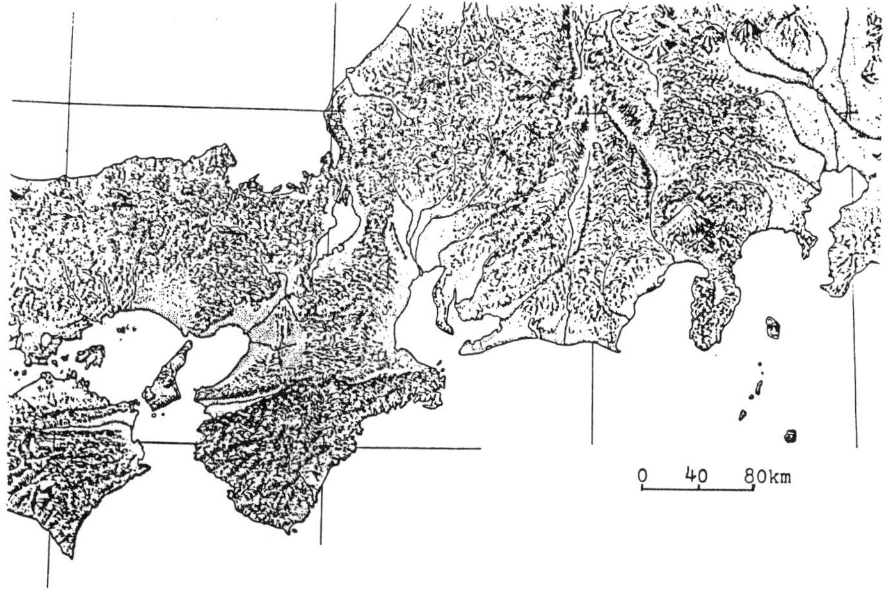

The great central cordillera of Japan, with its high ranges and narrow valleys, lies north and west of Mt. Fuji, centering on the Fuji, Tenryū and upper Kiso rivers and, to their north, the Nagano River watershed.

Source: Derived from foldout Physiographic Map by Guy-Harold Smith at p. 8 in Glenn Thomas Trewartha, *Japan: A Physical, Cultural, and Regional Geography*, 1st ed. (Madison: University of Wisconsin Press, 1963); courtesy of the American Geographical Society of New York.

Map 3. Central Japan: A Physiographic Map

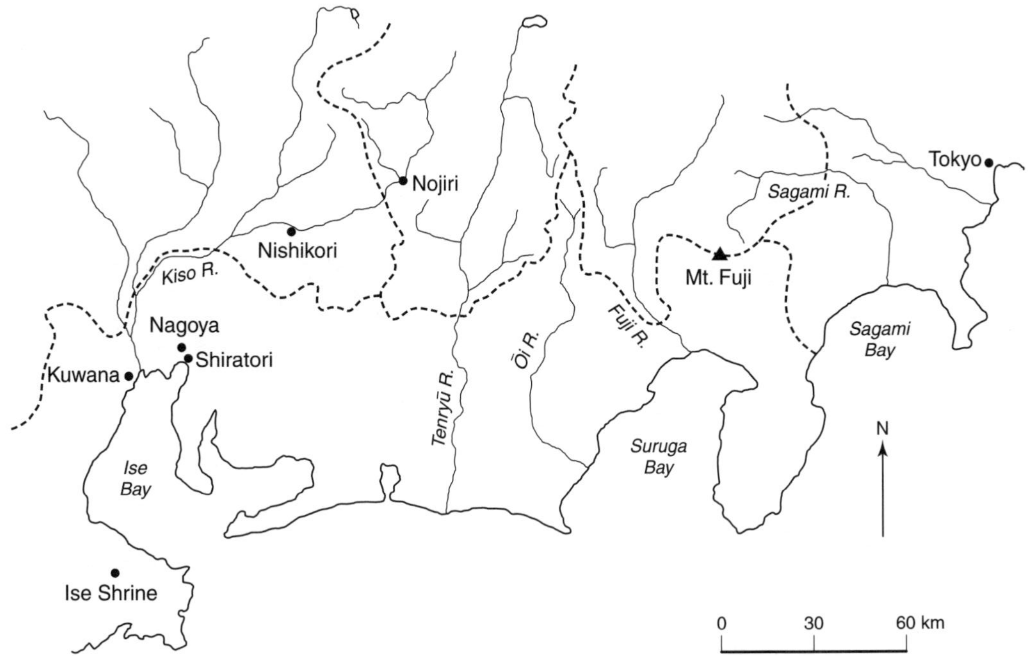

Map 4. The Tōkai Region

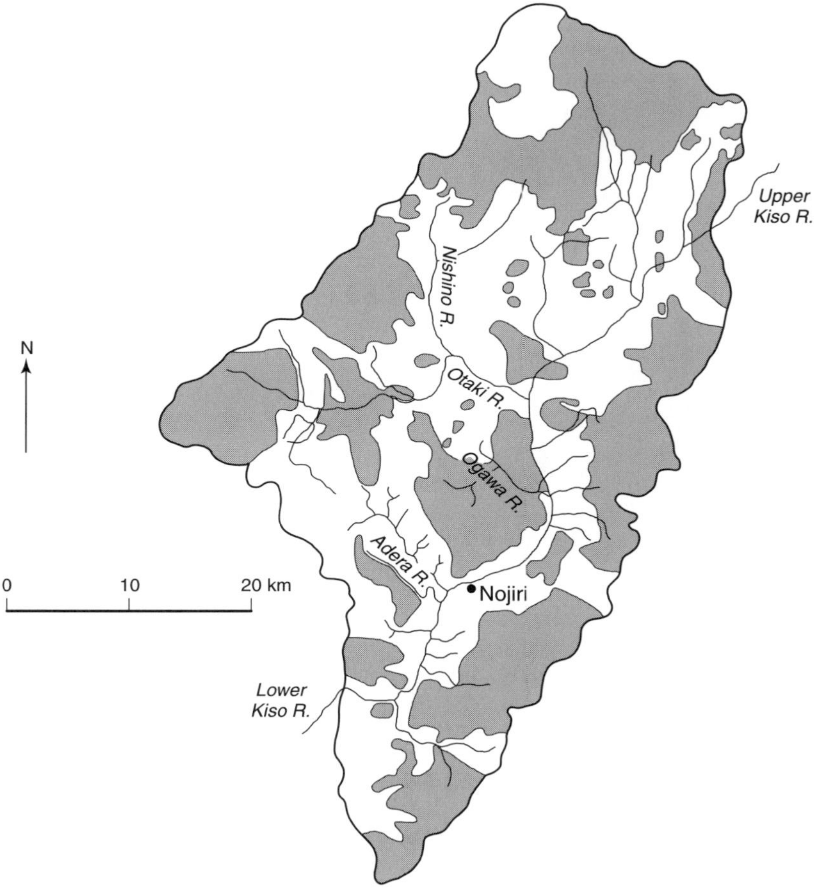

Legend:
dark lines: principal streams and tributaries
shaded areas: Goryōrin

One of the finest regions of Goryōrin was the Kiso River valley, where Imperial Forest actually extended beyond the valley, sprawling over ridge lines into adjoining watersheds and prefectures. As the map shows, however, even here Goryōrin was a severely convoluted set of properties. Goryōrin generally occupied more mountainous tracts, while the lowlands along rivers and streams were held as *min'yūrin*.

Source: The areas of Goryōrin are based on the map at Frontispiece of Machida Masami, *Kiso Goryōrin jiken* (Nagano-shi: Ginkawa shobō, 1982).

Map 5. The Kiso Valley (1905–16)

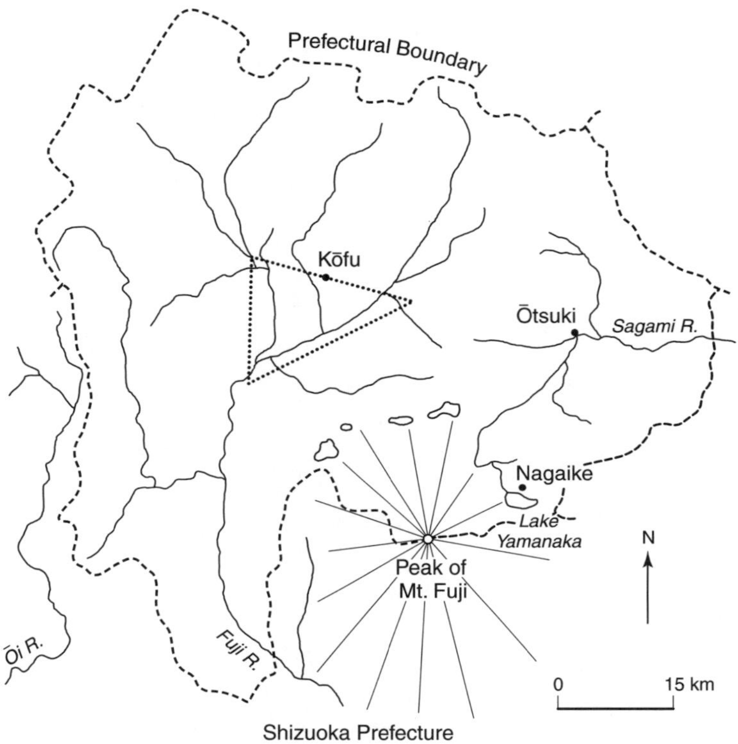

Mt. Fuji's lava skirt, marked by the radiating lines, extends northwards to the five small lakes (Fuji Goko) in southern Yamanshi. The Kōfu Basin spreads out south of the city proper. Only a few of the main tributaries of the Sagami and Fuji rivers are indicated.

Map 6. Yamanashi Prefecture

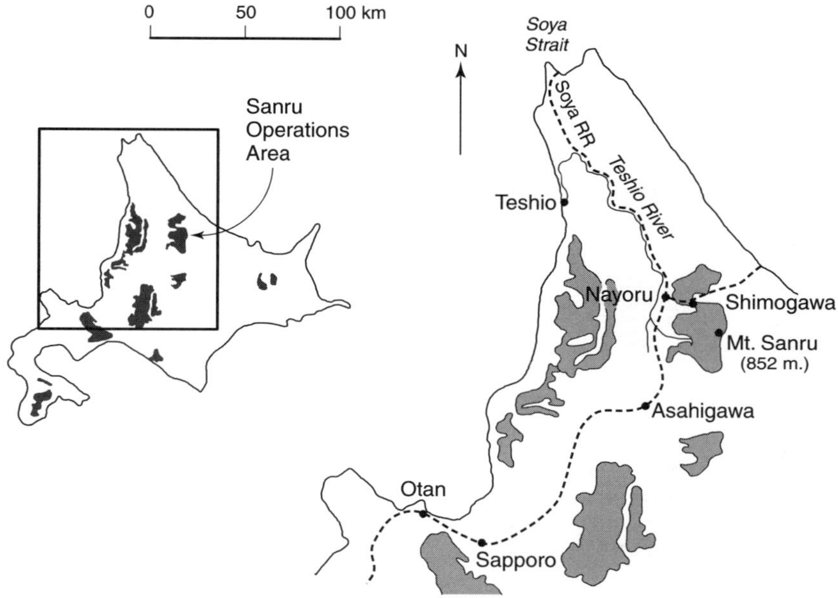

The comparatively large parcels of Imperial Forest in Hokkaido consisted of old-growth mixed forest with a substantial boreal conifer content. They were situated on moderately steep mountainsides that became more and more accessible as rail lines were opened.

Source: Goryōrin parcels based on the map at p. 258 of Kuroda Hisata, *Tennōke no zaisan* (Tokyo: San'ichi shobō, 1966).

Map 7. Goryōrin in Hokkaido (1937)

Preface
Britain and Japan: A Comparison

Observers of Japan often note the many ways in which it and Britain are similar – or dissimilar. A number of their points merit reiteration here because they bear directly or indirectly on the nearly untold story of Japan's "Imperial Forest" (Goryōrin), the analogue to Britain's Crown Lands or Royal Forest.[1]

SIMILARITIES

To note pertinent similarities first, Japan and Britain are north temperate-zone societies situated on comparably-sized island groups (see Map 1).[2] They both lie offshore the world's largest continental mass, and the biosystems of both reflect their locations, that of Japan being an expression of the East Asian temperate-forest biome; that of Britain, the West European. Similarly the societies and cultures of the two reflect the nearby continental civilizations. Both have been fundamentally shaped by those civilizations, but both also have retained assertively distinctive social and cultural qualities.

The two experienced roughly conterminous passage from forager cultures to early and more advanced agricultural societies. During their agricultural centuries both experienced substantial deforestation, which led to forest-product scarcities and substitutions and, from about the seventeenth century onward, to the development of reforestation strategies.

Both Britain and Japan were, by global standards, "early" industrializers, and today both have highly-urbanized, densely-settled populations, most of whose members enjoy comparatively high material standards of living. And like the rest of the industrialized world, both now depend on a global resource base for survival.

However, by their very nature as industrial societies, both contribute to the accelerating processes that are undermining the terrestrial and maritime biosystems on which that survival depends. These processes include global environmental pollution, resource depletion, climate change, declining biodiversity, and dissemination of exotica, including pathogens and parasites. Because of the great role forests play in determining global climate, air quality, biodiversity, and resource maintenance, the fate of global woodland – from the Siberian taiga to the Amazon watershed – is one of the most fundamental issues confronting the people of Britain and Japan. As a corollary, the condition of their own domestic forests becomes an issue of legitimate social and political concern.[3]

Finally, both Britain and Japan maintain storied monarchies that have played central roles in their histories and that stand today as perhaps the two most visible monarchies in the world. In both, those monarchies are at once a source of great pride and affection and a focus of irritation and resentment. For both, therefore, the nature of the monarch-populace relationship is a topic of continuing interest, which provokes both popular and academic curiosity about the monarchies and how they became what they are today.

DISSIMILARITIES

Fully as arresting as these – and other – similarities between Britain and Japan are the dissimilarities.[4] The more striking and pertinent ones can be examined in terms of the two societies' global positions, their geological history, and the effects of these variables on woodland and agriculture.

1. Global Position

Although both Britain and Japan are temperate-zone societies, the northernmost point in Japan falls along a latitude nearly 500 kilometers south of the southern tip of Cornwall. And yet, because Britain sits on the eastern side of the Atlantic while Japan is perched on the eastern edge of Eurasia, seasonal flux in the trajectory of prevailing westerly winds makes Japan much the colder of the two in winter.

In Japan, substantial accumulations of snow last for months, especially at higher elevations in the country's northern half. In the British Isles, by contrast, the prevailing Atlantic westerlies mostly bring cold rain and dampness during winter. This weather pattern enables peat moss to flourish, with the result that, in Britain, denuded forests are very susceptible to takeover by the moss, which can then stifle natural reforestation if abetted by continuing human removal of forest canopy, whether for fuel or other use. Japan's woodland, being spared this danger, finds it easier to recover from overuse.

That moss, however, does in time build up appreciable depths of peat, which can be used as smoky but serviceable fuel in lieu of the lost wood and charcoal. In effect, its use conditions a populace to look downward rather than upward for fuel supplies and also prepares it for the rather more difficult task of extracting and burning ground coal. So it is not surprising, perhaps, that in Britain one does see an autochthonous succession from use of fuel wood to peat to ground coal. In Japan, on the other hand, the leap directly from fuel wood and charcoal to ground coal developed later, during the eighteenth century and beyond. And when it did occur, it generated more public objections, which substantially slowed the diffusion of coal use until exogenous factors overwhelmed the resistance in the 1860s.

2. Geological History

At least as fundamental as global position in shaping the destinies of Britain and Japan were their geological origins and development. Those are strikingly dissimilar.[5] The British Isles are a slightly migrant piece of ancient tectonic plate that initially formed some 400 to 300 million years ago when slowly moving continental plates collided to create the super continent, Pangaea. After Pangaea disassembled some 200 to 100 million years ago, the vicinity of today's Isles survived as a marginal part of the Eurasian Plate. However, during the most recent sixty five million years or so, as that giant plate gradually rotated in a clockwise direction, the Isles drifted northward, becoming separated from their motherland. That separation produced the shallow English Channel and North Sea, giving Britain the offshore location that has so influenced its human history.

Japan, by contrast, is a mere infant, a child of agonizing birth. It was driven up out of the bowels of the ocean during the past fifteen million years – and mostly during the past two-five million – by the ongoing collision of four plates, two continental and two oceanic. The eastern edge of the rotating Eurasian Plate, driving in a southeasterly direction, encountered the westward-bound North American Plate, locking onto its leading edge and steering it southward. As the two plates descended, they met two oceanic plates, the northward-moving Philippine Plate and the northwestward-headed Pacific Plate. Where the four collided, they slowly, grindingly produced – and are still producing, as every earthquake and volcanic eruption reminds us – the Japanese archipelago.[6]

So, the British Isles have an extremely archaic provenance and in consequence are well worn down, being dominated by lowlands and gently-undulating, low hills. Even the celebrated Highlands of the north almost all stand under 1,000 meters in height and are well rounded with age. In striking contrast, Japan has a very recent and brutal provenance that has left it extremely mountainous.

Indeed, about 80% of Japan is steep hillside and upland. Not only are ridgelines acutely upthrust and often razor-sharp, but also they are high. Entire mountain spines stand higher than Scotland's grandest summits. Innumerable peaks tower above Ben Nevis's 1,343 meters, with many standing in the 2,000 to 3,000 meter range, volcanic Fuji-san the tallest at 3,776 meters. Sandwiched between the towering ridgelines are elaborate dendritic valley systems down which small streams plunge. During heavy rains and rapid snow melt, these can suddenly turn into raging torrents, carrying debris of every sort, depositing it on the slender alluvial lowlands and small coastal plains that border the mountain ranges and front the ocean.

These topographical dissimilarities between Britain and Japan were heightened by weather patterns of the Pleistocene Ice Age. The vast glaciers that periodically entombed the British Isles until some 15,000 years ago accelerated the processes of erosion which lowered and rounded hills, moving detritus to the lowlands where it was more amenable to further weathering and evolution into fertile soil. Japan, by contrast, acquired only a few small mountaintop glaciers, so its steep hillsides have remained intact down to the present. Their shallow rocky soil has survived unscoured, held in place by existing vegetation, but subject to easy dislocation by earthquake, wildfire, or human abuse.

3. Effects on Agriculture and Woodland

Thanks in great part to their particular combinations of global position and geological history, the two realms have had very dissimilar agricultural experiences and ended up with strikingly dissimilar forest conditions.

The agricultural experiences of the two were, of course, influenced in the first instance by Britain's exposure to the dry field/pasture practices of continental Europe and Japan's exposure to the dry field/wet rice cultures that had arisen in East and Southeast Asia. However, in the absence of favorable topography and climate, things would have worked out differently in both cases. The geological heritage of the British Isles left them with extensive areas fit for pasturage and dry-field tillage. In Japan, by contrast, most land was too steep for either use, and much of the potentially arable 20% was flat alluvial deposits amenable to irrigation.

The climates of the two then reinforced this difference. Japan's hot summers, the product of ocean-warmed southwesterly winds (in contrast to winter winds from Siberia), permitted irrigated rice culture, while Britain's moist but cooler summers favored dry grains and pasture. Because wet-rice cropping has, from the beginning, yielded a greater caloric count per hectare than either pasturage or dry grain culture, more and more of Japan's tillable acreage was converted to irrigated fields as time passed. This choice of harvests compensated for the archipelago's paucity of arable land, enabling it to support a human population that has for centuries been roughly twice that of the British Isles.[7]

As is implicit in those differences of agricultural experience, Japan and Britain today have very dissimilar domestic woodland situations. Japan remains a heavily-forested realm; indeed, some 70% of its land is still wooded to one or another degree.[8] In Britain, by contrast, woodland is restricted to a small portion of the total land surface. Over the centuries most acreage has been converted to pasture or tillage in lowland areas or lost to peat moss and heather in others, so that during the 1960s woodland in the United Kingdom was recorded at 4,305,100 acres, only about 7% of the total land area.[9]

REGARDING WOODLAND

Woodland being the focus of this study, let us look a bit more closely at the forest biomes of Britain and Japan. Both realms are situated on the margins of the Eurasian continental temperate-forest zone, but they are at opposite margins, and that placement has yielded basic and important silvicultural dissimilarities.

The main point is that Japan has been able to partake fully of the biological riches found in the East Asian biome. That biome is so lush in the first instance because the high Himalayas turn southward and peter out many hundreds of kilometers west of the South China Sea. Consequently there are no major mountain chains separating East and Southeast Asia, which has expedited vegetational mingling and migration between the two. In addition, the East Asian region was not scoured or buried by Ice Age glaciers, so it was able to preserve into recent millennia its exceptionally diverse and lush biosystem. And Japan has been able to participate fully in that botanical richness because of prevailing winds, ocean currents, and – most importantly, perhaps – Ice-Age-related sea-level shifts that periodically linked the archipelago to the continent for long stretches of time. In consequence, Japan's mountains have remained sheathed in a wonderfully complex and adaptable vegetative cover.

By comparison, the indigenous biotic heritage of the British Isles is thin. The glaciers that covered them and extended over northwestern continental Europe, together with the adjoining tundra belt that covered much of the remaining area north of the Pyrenees and Alps, devastated the existing vegetative cover. Then the Mediterranean Sea and its associated mountain chains obstructed the plant migration from more southerly latitudes that gradually reforested the region after the glaciers receded some 15–13,000 years ago. The Channel and North Sea remained exposed as dry land long enough for such vegetation as was reappearing in northwest Europe to re-establish itself in the Isles. But even so, that ice-age legacy left Britain during recent millennia with one of the world's least-diverse temperate-forest biomes.[10]

These dissimilarities in post-glacial woodland circumstance contributed to differences in the way British and Japanese forest history developed. In both places the spread of agriculture, its subsequent intensification, and the population growth that accompanied those trends led to extensive deforestation and increased demand on the woodland that survived. In Britain, as noted above, the permanent displacement of forest was nearly complete. In Japan, much less terrain was converted to arable and very little to pasture. But population pressure sustained widespread and risky Japanese attempts to exploit steep hillsides, commonly as swidden. That slash-and-burn activity, together with intensive harvesting of brushland for fuel, fodder, and mulch, caused chronic problems of erosion and flooding. And over the centuries it transformed many hillsides into poorly-wooded ranges of scrub pine, tropical grasses (diverse bamboos), forbs, vines, and other low growth, while some areas degenerated into barren hillside.

In both Britain and Japan, then, well-wooded forest gradually was lost, and eventually that trend provoked responses that were similar in some ways, dissimilar in others. Britain, as noted above, experienced a fuel-supply transition from wood to peat to ground coal that Japan underwent centuries later and without the peat interregnum. To supplement construction timber, moreover, both societies employed mud in the form of plaster and tile. But Britain, unlike Japan, also adopted brick construction and masonry techniques that utilized cementing material.

Both societies developed elaborate systems of land-use rights intended to regulate the exploitation of woodland. And both began trying to conserve woodland stands and then to nurture new ones. However, these latter trends appeared roughly a century sooner in Britain (where they date from the later sixteenth century) than in Japan. Both introduced exotics to enhance timber production, but whereas the practice began in Britain around 1550 with the importing of Norway Spruce, in Japan it only began around 1890 with the importing of German Spruce. Finally, the most striking dissimilarity may have been the early British resort to timber imports – from the Baltic littoral, North America, and India in particular – from the later sixteenth century onward, whereas Japan did not begin obtaining wood from abroad until the twentieth century. By late in the century, however, both societies had become heavily dependent on imported wood products, and they still are.

REGARDING MONARCHY AND ITS LANDED PROPERTY

Britain and Japan have two of the world's most celebrated royal traditions. Yet the two are dissimilar in many ways, including their relationship to the land, which is our main interest here.

Britain has had a succession of monarchial lineages that have risen and fallen through vigorous participation in political life. Dominating other power holders when possible, striking political bargains of convenience as necessary, and warring when required to protect their position, they long played a role of *primus inter pares*, at least until recent centuries witnessed the emergence of dominant kingship and, in the industrial age, titular royal sovereignty. In Japan, by contrast, a single, loosely-defined royal lineage, the "Imperial Household," has retained its status for some 1,500 years. However, for the last 800 years or so, it has been largely titular, with real power in the hands of others, who then used ties to that Household to sanction their exercise of authority.

The Imperial Household's loss of real power meant, of course, that it also lost control of land. Ancient emperors claimed to possess all the realm, and around AD 700 they still played a major role in settling issues of land control throughout their empire. By AD 1500, however, their descendants retained only a few small parcels that provided them with food, fuel, and other necessities. And they held even those thanks largely to bargains struck with ambitious power holders, a situation that continued until the 1880s. Then, when the Imperial Household did acquire large tracts of woodland – over 2,000,000 hectares – around 1890, these were administered independently and regarded as distinct from both government land and land held as "private property" (by villagers, entrepreneurial lumbermen, communal groups, municipalities, and religious bodies).

In Britain, meanwhile, although ambitious early kings may have claimed the realm as their own, more and more of the "royal forest" – which originally denoted such land as the monarch wished reserved for his hunt – was being converted to permanent arable and pasturage under local control. Moreover, political needs of the moment led kings gradually to award more and more wooded acreage to their allies as estate land. And because monarchs were active heads of government, the remaining Crown Lands came increasingly to function as government land.

During the reign of George III (1760–1820), even as programs of forest rehabilitation were beginning to show results, Crown Lands became government property in all but name, their revenues turned over to the Exchequer, their management to a designated government agency. And by the nineteenth to twentieth centuries the groups concerned about forest recovery – epitomized by the Royal Scottish Forest Society (dating from 1854), the Royal Forestry Society (from 1880), and the Forestry Commission (from 1919) – tended to look beyond matters of ownership (whether as Crown land, large estate, or small-holder woods) to the broader issue of timber production, which was the dominant concern of forest improvers.[11]

In short, by the time our story of Goryōrin commences around 1890, royal forest in Britain was an anachronistic concept. Crown Lands totaled only about 115,000 hectares, or 6–7% as much land as Japan's Imperial Forest. Moreover, they were government property and were being handled as part of the larger forest biome within which they existed.

In Japan, by way of contrast, a long-marginal Imperial Household was being newly rehabilitated, endowed with sovereignty, equipped with its own organization and material resources, and elevated to the status of major woodland holder. Its widely-scattered properties were clearly distinguished from both government woodland and the myriad small parcels of private property. They were administered independently and were viewed as a crucial element in the funding and preservation of that sovereign Household. Let us try to find out why that was so and how matters worked out.

Introduction

Before World War II, some five to six per cent of Japan's woodlands were known as Imperial Forest (Goryōrin).[1] They were under the administrative control of the Imperial Household's Property Office in Tokyo (Goryōkyoku; later called Teishitsu rin'yakyoku) and its many branch offices throughout the country.[2] The story of why and how Goryōrin came to be, of the vicissitudes it experienced, of how it was managed, how and why that management changed as time passed, and how well or poorly the whole enterprise worked out – these several topics are the foci of this study.

The study, then, is basically an inquiry into silvicultural history, broadly defined. In Chapters 1 to 3 we examine the antecedents and origins of Goryōrin, the logic of its creation, and the overall pattern of its evolution. Then, after looking more closely in Chapter 4 at the painful experience of Goryōrin in Yamanashi Prefecture, we focus in Chapters 5 and 6 on more technical aspects of forest operations as they developed and changed during the fifty-eight-year life of this woodland system. In Chapter 7 we recapitulate the main themes of our story, offer some assessments of the Goryōrin experience, and take a final look at it in comparison to Britain's Crown Lands.

In Part II (Chapter 8), by way of further background in our understanding of the wider implications of the implementation of Goryōrin, I examine the major restructuring policy of the Meiji reformers in the "Kan/Min" division of woodland throughout Japan, which created the greatest conflict between government and people in the history of the country. This chapter is written and refenced as a stand-alone essay for those readers who wish to examine the *kan/min* topic independently of the subsequent Goryōrin experience.

In the end, however, history is an infrangible whole. So as we proceed, we shall find ourselves touching a number of themes that have appeared prominently in numerous studies of Japanese history, and, in one form or another, in historical studies more broadly.

As suggested in the Preface, to begin with, geography mattered. Specifically, Japan's highly-mountainous character exercised a major role in shaping this history. Long before 1868, woodland had become restricted almost entirely to hillsides, with the lowlands devoted to agriculture and human habitation. And since most of the accessible woodland, on which villagers depended for survival, was on hillsides directly above villages and farmland, the centuries of human use had long since created widespread problems of – and yet greater potential for – erosion, landslide, and flooding, with all their social ramifications. The question of who could best manage those hills so as to avoid such problems while benefiting from their yield was, as we shall see, one source of the chronic friction between villagers and the makers of Goryōrin that played a major role in its history.[3]

Similarly, a truism we all recognize is that the past shapes the present. The behavior of all of us is influenced by our communal heritage, whether at the family level or that of larger-scale groupings. Chapter 1 explores most fully the facets of Japan's heritage – the legacy of imperial rule, patterns of land control and utilization that had developed over the centuries and survived into the Meiji period, and the character and problems of the ruling group that seized power in 1868 – that shaped the formation of Goryōrin, its character, and evolution.

Another universal truism is that social groups interact, shaping one another's experiences. Scholars of Japan seem particularly interested in how "foreign" and "domestic" factors – politico-economic interests, technology, ideas, institutional arrangements – have interplayed to shape Japan's modern experience. Chapter 2 addresses that theme most fully, but it recurs from time to time thereafter.

Equally universal is the way in which various issues of the moment interact, shaping one another's outcomes. We shall continually note extraneous political, diplomatic, economic, social, and environmental considerations that played a role in shaping the Goryōrin experience from beginning to end.

Another pair of truisms that historians (and polemicists) have noted repeatedly is that vast imbalances of power and privilege are endemic to the human condition and that those atop the social heap do their best to stay there. Those conditions show up in the story of Goryōrin, most notably as a pervasive and persistent tension between the creators and operators of Goryōrin on the one hand and, on the other, the villagers whose local woodland-use arrangements and opportunities were threatened or injured by their efforts. Conflicts over the control and use of this woodland show up throughout our story, but especially in Chapters 3 and 4.

All these themes will catch our attention from time to time. But at the heart of our story are two topics: Japan's modern monarchy and its modern forest system.

The Japanese monarchy has been treated extensively, and for the decades covered by this study, a couple of simple but perdurable stereotypes have been prominent in the literature. One is the benign image of imperial virtue that Japan's leaders promoted before World War II and that the child in all of us would dearly love to see expressed in verdant forests cherished by loyal and well-loved subjects. The other, contrarily, is the malign image of a ruthless and irresistible imperial authoritarianism that some envisage as running roughshod over the well-being of its own subjects, as well as anyone else who got in its way, at least until its comeuppance in 1945.[4] The Goryōrin experience illuminates both the rise of the modern Imperial Household and key aspects of its fiscal system. But the picture that emerges is one of complexities and ambiguities that suggest a far less clearcut role for the monarchy, a role laced with mixed motives, effects, and attainments.

So observers of monarchy may find things of interest in this study. However, it speaks more centrally to students of forest history and silviculture. For several reasons it sheds light not just on the experience of Goryōrin but also on the broader story of Japan's government forest (*kokuyūrin*) during the seventy-odd years to 1945 (see Table Int-1).[5]

Specifically, although *kokuyūrin* dates from shortly after the Meiji Restoration of 1868 and the smaller Goryōrin from three decades later, the latter was essentially a diminutive version of the former. It was composed of the same sorts of land and vegetation and was created to serve similar purposes. It was organized and managed in much the same way by people with similar training and values. It was linked to the broader society in almost the same fashion. And finally, it was plagued by the same sorts of problems, which elicited nearly the same remedies and resulted in a generally similar outcome.

So the history of Goryōrin stands as a microcosm of broader conditions in and around Japan's forests during the decades from Restoration to military disaster. It shows how rulers labored to transform the variegated woodland realm they had inherited from the Tokugawa into a consolidated one whose production would help preserve their regime and sustain a vigorous industrial state and society. It gives us a few glimpses of villagers struggling to cope with disruptive changes of the day and

Table Int-1: Forest Acreage and Stands, 1930s

Category	*Acreage* (1000 *chōbu*)	% of total	*Stumpage* (million *koku*)	% of total	*Koku* per *chōbu*
Goryōrin	1,335	5.5	562	8.4	421
Kokuyūrin	7,650	31.7	2,999	44.9	392
Kōyūrin	4,431	18.4	813	12.2	183
Shajirin	152	0.6	40	0.6	263
Min'yūrin	10,512	43.6	2,255	33.8	214
Total	24,080	100.0	6,668	100.0	277

(Source: Kuroda Hisata, *Tennōke no zaisan,* p. 112. Acreage in 1939; stumpage in 1934–35. "*Koku* per *chōbu*" figures added here. *Kōyūrin* refers to communal properties; *shajirin* to fane properties, and *min'yūrin* to other forms of "private" property. Commonly *kōyūrin* and *shajirin* are treated (as in Table E-1, Appendix E) as a part of *min'yūrin*.)

merchants maneuvering to seize the moment. It reveals how and why foresters attempted to enhance particular types of forest output, utilizing practices inherited from Japan's own silvicultural legacy and modifying them with new ideas and technologies from abroad as they strove to cope with new demands, circumstances, and expectations. And in doing these things, it helps us to look forward, beyond the 1940s, by giving us a foundation that aids our understanding of how and why Japan's forests could undergo the striking changes they have experienced since World War II and Goryōrin's dissolution.[6]

PART I

1

Historical Background

Goryōrin was a modern invention, a Meiji-era solution to problems of the day. Like so many Meiji solutions, of course, it was grounded in earlier precedents, both domestic and foreign. However, those precedents were full of mutual inconsistencies, and many were ill-fitted to the times. So we shall see, as our story unfolds, that both the devising and implementing of policy on imperial land-holding were riddled with uncertainty and disagreement and involved much trial and error.

PRE-MEIJI IMPERIAL LANDS

The insurgent samurai who in 1868 toppled the Tokugawa regime and formed their own, the Meiji government, claimed to have restored ruling authority to the emperor. It was a rationale that helped justify not only their "rebellion" against the shogun, liege lord of their own lords (*daimyō*), but also their assaults upon those other daimyo who refused to "betray" the shogun. The immediate convenience of their rhetoric notwithstanding, however, these restorationist claims, like political rationales in general, carried more complications than their proponents seem to have realized at the time.

Central to the claim of "restoration" was a principle that enabled the insurgents to assert their "right" to assign and reassign land as they saw fit and to extract from it such usufruct as they deemed appropriate. The principle had been nicely expressed some 1100 years earlier by Emperor Shōmu (r. 724–49). In the year 743, when dedicating a grand new Buddhist statue at Tōdaiji in Nara, he declared:

> It is We who possess the wealth of the land; it is We who possess all power in the land.[1]

And in his day all the realm did "belong" – nominally – to the emperor. His government allocated parcels of land to deserving officials and institutions and even to cultivators. And recipients were expected, in grateful return, to provide loyal service and material tribute as specified by law.

In later generations, the tribute-collecting power of the imperial court waned, prompting its eleventh- and twelfth-century leaders to sequester more and more parcels of land in their own "private" estates, the imperial *shōen.* Those they held in the name of one or another retired emperor, imperial consort, or branch lineage. The arable parts of their *shōen* generated foodstuffs for the Imperial Household while forested portions provided firewood, charcoal, and construction material, along with diverse other consumables and barter goods.

During the fifteenth and sixteenth centuries, as civil war engulfed the realm, nearly all these imperial *shōen* were seized by ambitious figures, notably samurai chiefs and religious leaders, leaving the court an impoverished shadow of its former self. A few villages in the Kyoto vicinity did continue to provide goods and services to the emperor, including fuel and building materials that came from hamlets nestled in the nearby valleys of the Takano, Kamo, Kiyotaki, and upper Ōi rivers. For the most part, however, the court had become dependent on "the kindness of strangers," assistance from those who hoped to benefit from displays of solicitude for the August Personage.

Powerful military leaders pacified the realm between about 1570 and 1620, and to help legitimize their dominion, they restored some semblance of imperial dignity. They reactivated courtly rituals, repaired imperial tombs, built new palaces and villas, and recognized these structures and the land on which they stood as imperial property (*kinri goryōchi*). The court also retained its right to obtain wood and other goods from several nearby villages, whose members took some pride in their longstanding role as purveyors to the throne.

To assure the emperor's basic well-being, moreover, after 1600 the newly-triumphant Tokugawa shogun assigned him the yield from specified agricultural properties. By the early 1700s, these several properties – which were overseen by Tokugawa appointees, not those of the court – furnished about half the Imperial Household's total revenues. For all their necessity to the throne, however, they constituted only about one tenth of one percent of Japan's total estimated agricultural terrain.[2] The court was thus the recipient, not the assigner, of land, and what it received was a barely consequent part of the realm. Shōmu would not have been impressed.

EMPEROR AND GOVERNMENT IN EARLY MEIJI

Those who overthrew the Tokugawa in 1868 purported to be re-establishing the emperor's governing power and authority, and one of their first major steps in that project was to claim as imperial domain all lands of the Tokugawa regime (shogunate or *bakufu*) – about one-seventh of Japan's total. This they combined with the existing parcels of *kinri goryōchi* to form the nucleus of the imperial realm they were reconstituting.

The taxable yield of this realm was to flow into a central treasury, which would then allocate funds to the Imperial Household Ministry

(Kunaishō, created in July 1869; today's Kunaichō). These funds were to cover expenses of the emperor, his family, and household staff. Over the next couple of years the new regime claimed all the remaining lands of Japan, specifically those of the hereditary civil barons (daimyō and *hatamoto*) and the fanes (*shaji*: Shintō shrines and Buddhist temples). By the end of 1871 the process was complete; the nineteen-year-old emperor again "possessed" the realm. But since his household allowance came via the central treasury, which neither he nor his household controlled, he also remained as completely at the fiscal mercy of others as during the preceding 450 years.

These "others," the early Meiji leaders, were a motley crew. Mostly middling samurai from the daimyo domains of Satsuma and Chōshū in southwest Japan, they also included a handful of court nobles, daimyo, and samurai from other baronies. And on their periphery stood a modest assortment of collaborating merchants, village leaders, and religious figures.

Even as the new rulers secured their positions vis-à-vis Tokugawa holdouts and other challengers, they became embroiled in internal rivalries and disagreements. Among the many divisive issues of the day, one point of contention sprang from the fact that their regime was far more than a mere imperial institution. Somehow they had to distinguish between the government as a whole and the Imperial Household as a discrete entity. How precisely to fit the two together – the treasury/kunaishō arrangement of 1869 was one early effort – became an issue of hot debate that involved questions of political ideology as well as power and privilege.[3] Amidst arguments about governing principle and precedent, both Japanese and European, leaders pursued a twenty-year-long process of trial and error that slowly rearranged fiscal, administrative, and ritual practice. By the 1890s, they had finally established what they considered a reasonably workable relationship between Household and regime.[4]

In terms of land, which concerns us here, in 1872 Meiji leaders recognized "Imperial Household Land" (called for a time *kōgūchi* rather than *goryōchi*) as a distinct sub-category within the broader category of "government land" (*kan'yūchi*). This Imperial Household Land consisted primarily of palace, villa, and tomb sites, most of it in Tokyo and Kyoto, and as of 1880 it totaled only 634 *chōbu* (629 hectares; approx. 1600 acres).[5] The emperor thus had land, but prior to the 1880s it was not income-producing acreage. And his household finances continued to be handled as part of the national fisc, subject to the same political pressures as all other parts of the budget.

FOREST ISSUES IN MEIJI LAND REFORM

Questions of imperial authority, organization, and ritual generated conflict among the new rulers and their close allies. Questions relating to land control generated vastly more widespread, intense, and destructive conflict that mainly pitted the new regime against the country's village population.[6] That conflict mostly involved woodland, and it persisted for decades. So we must ask why the rulers were even interested in land; why woodland in particular presented them with such thorny problems; and how the problems were handled.

As children of Tokugawa society, early Meiji leaders viewed land as either arable or other. Arable was what really concerned them at first because they viewed it as the main source of land-tax income. They generally agreed that arable should be handled by farmers who would then pay taxes on it. Other land, however, had diverse uses, and much of it also presented dangers because of its deteriorated condition. So, how – and by whom – that land should be handled was not at all clear. And as leaders explored various choices during the seventies, they discovered that all were fraught with complexity.

Settling Arable Land

Establishing policy for control of arable land was relatively straightforward because records of prior use were extensive and conflicting claims relatively few. Complexities did exist, however, because in the Tokugawa system that Meiji reformers had inherited, some 300 daimyo controlled nearly three quarters of the realm. Each of these lords, following his own rules and using his own staff, collected tribute for his own regime from the villages within his domain. By and large the amount he extracted from a taxpayer was based on the payer's ascribed "use rights" (*riyōken* or *yōekiken* in modern parlance), primarily as recorded in village registers that specified the putative agricultural output (*kokudaka*) of parcels of arable land whose tillers were listed by name. Particulars of the system varied greatly from domain to domain, and Meiji leaders sought to replace this fragmented arrangement with a nationally unified system of taxable "property rights" or "ownership" (*shoyūken*), as found in contemporary Western Europe.[7]

The reformers found a starting point for their project of replacing use rights with ownership in their assertion that all the realm was imperial land. On that basis they claimed that the emperor's government had the right to determine all land ownership. As a first step it must implement a basic "*kan/min* division" (*kanminyū kubetsu*) that would clearly distinguish tax-exempt "government land" (*kan'yūchi*) from taxable "people's land" (*min'yūchi*), all parcels of the latter to be identified with specific owners. To implement that division, it was the "people," not the regime, who had to present evidence in support of their claims to "ownership" of specific parcels. Once land claimants had done that to the satisfaction of government examiners, assessors could place a "market value" (however determined) on each parcel of *min'yūchi*, and the government could then tax it on the basis of that valuation.[8] All other areas would remain tax-free government land.

On arable land that task was largely finished by 1877. However, the *kan/min* differentiation of woodland proved to be a much larger, immensely more complex, and, consequently, long-drawn-out undertaking. Indeed, the process generated such complications that in 1881, the government attempted to mollify public discontent and get on with other tasks by declaring tax reform officially completed. In fact, however, the job was far from finished, and the determining of woodland rights and boundaries continued for decades thereafter, contributing to Goryōrin's difficulties. So the topic requires closer scrutiny.

The Problem of Woodland

Sorting out woodland was a large and thorny enterprise for several reasons. One problem was that whereas reformers agreed on what they wanted arable land to do, for woodland they envisaged several purposes, not all of them mutually compatible. Most notably they wanted to control erosion and flooding; they also wanted timber and other yield; and they wanted income. So they wanted to place some woodland under government control as either protection forest or production forest while selling some to entrepreneurs. But how to proceed and how much land and which specific parcels to treat one way or another were recurrent sources of intense internal disagreement.

Kan/min division was such a large task on woodland because upwards of eighty per cent of Japan's land surface was classified as *sanrin gen'ya* (woodland and waste), as distinct from arable.[9] And most of that land was steep hillside that had to be surveyed. The sheer difficulty of trudging up and down those densely thicketed hillsides to determine boundaries, acreage, and stand conditions was an exhausting, time-consuming, error-prone task.

That surveying was required, however, because even where woodland was reasonably accessible, very little had ever been measured with any degree of accuracy. So most records were unclear or worse. When new surveys were made and old truths were challenged or found to be in error, the discoveries often disrupted local arrangements to someone's disadvantage. And predictably, such disruptions could and did lead to disputes that slowed, complicated, and sometimes utterly stalled the process of *kan/min* division.

Changes in the control and use of woodland were particularly disruptive because the existing use-rights arrangements there were vastly more complicated than those on arable. A user of tillable land generally had an exclusive and annually recurring seasonal right to cultivate and harvest specified fields. In contrast, much woodland, and especially the more accessible and hence more useful parts, consisted of mixed perennial vegetation that had multiple uses. Throughout the realm such land was generally managed by one or another type of communal (*iriai*) or rental arrangement that authorized many specified users but might spell out in great detail precisely who could do what, where, when, how, and how much. So, whereas the resolution of conflicting claims to arable generally produced one winner and one loser, in woodland any attempt to designate owners and terminate the use rights of others was very likely to produce several losers for every winner. Unsurprisingly, the level of resistance matched the level of provocation.

Figs 1-1a, 1-1b. *Sanrin gen'ya. Sanrin,* or "mountain forest," (left above), refers to well-wooded steep hillside, which during the twentieth century consisted more and more of plantation stands of *sugi* and *hinoki* (see also Fig. 2-1). *Gen'ya,* or "wasteland" (right above) generally was found on lower hills and consisted of mixed growth, mostly underbrush, grasses, and forbs, with little timber content (see also Fig. 3-1). Mingled in it might be patches of swidden, arable, or small plantation stands. Being subject to sustained use by villagers, *gen'ya* generally was close to hamlets. (From author's collection.)

Dealing with Woodland Problems

During the 1870s, land reformers tried to cope with this thicket of difficulties by devising compromise mechanisms that perpetuated de facto *iriai* practice and accommodated rental agreements. Specifically, in 1872, after the initial moves to establish private ownership of communal land had elicited a storm of hinterland protest, they created a category of "public-use land" (*kōyūchi*) for parcels claimed by local groups and villages. As the seventies advanced, they repeatedly tinkered with the notion of *kōyūchi,* sometimes to quiet discontent but sometimes to advance their own agenda. And they tried to reduce conflict over parcels designated as *kan'yū* and *min'yū* land by establishing license and fee systems that more or less replicated older use-rights and rental arrangements within a nominal framework of property ownership.

By creating these devices, the reformers quieted enough of the hinterland outrage so they could press ahead with their land surveying and

Fig. 1-2. Uses of *gen'ya*. Some of the diverse uses of *gen'ya* are suggested by this snapshot, from southwest of Tokyo, ca. 1980. The oak coppice has been cut to produce three-foot sticks that will be stacked in the woods to grow *shiitake* (mushrooms). Smaller sticks in the background can be used for farm construction or fuelwood, and the bundled twigs can serve as kindling, mulch, or fencing material. Dwarf bamboo (*sasa*), which fills the space between trees, is cut annually for fodder and mulch. (From author's collection.)

categorizing. But they also compromised the principle of ownership and obfuscated the distinction between *min'yū* and *kan'yū* property.

That outcome was frustrating to the reformers, prompting them in later decades to nibble away at the concessions. Nor did the compromises really satisfy villagers. For one thing, compared to the old arrangements these substitute mechanisms were cumbersome and constrictive in various ways. Also, they generally were more costly to the user since the generating of government income had been one objective in their creation. Moreover, they sacrificed innumerable local idiosyncracies to achieve national uniformity of practice. Most irksome of all, probably, they vested active legal control in outsiders rather than in the locals who had operated the older multiple-use arrangements.

Because of these complications, the task of surveying woodland went on for years, in many areas being repeated once, twice, or even thrice before authorities were generally satisfied during the 1920s. And the process of establishing ownership and working out acceptable relations with people of the locality proved so tortuous and conflict-ridden that it plagued authorities and people for decades. Finally, and perhaps worst of all, because *kan/min* division so often presented villagers with a "use it or lose it" choice, it provoked a rash of incidents of reckless harvesting and woodland damage (see Table 8-1 in Chapter 8). Mostly these occurred during the early years, but some continued on into the twentieth century, even until after World War II in places where old grudges had acquired a life of their own.

* * * * *

Long before these woodland issues even began to be resolved, the matter of Goryōrin made its appearance. The disruption, turmoil, and opportunities of early Meiji had given rise to energetic political activism, a host of local protests and uprisings, and during 1877 the Satsuma Rebellion, a major armed challenge that the new regime suppressed only with great difficulty and at ruinous financial cost. The political activism was largely an expression of personal ambition and rivalries that drew strength from the widespread hurts and hardships of the day. But at an ideological level it embraced a noisy rhetoric about "freedom and people's rights" (*jiyū minken*), and even republicanism. And this rhetoric contained a usually implicit but occasionally explicit questioning of the need for an imperial institution itself. Active interest in the formation of a large imperial estate appeared around 1880, primarily as a response to that rhetoric.

2

Establishing Goryōrin

Meiji leaders saw potential danger for the imperial institution, and hence for their regime, in both the political rhetoric and circumstances of the 1870s-1880s. That perception led them to ask how best to counter the danger, in particular how to assure that the institution would never again become a financial hostage to others, whether feckless politicians or a fickle public. During the later 1870s, answers to that question began to appear, and during the eighties they were clarified and implemented.

As the 1880s advanced, however, it gradually became apparent that the first general strategy – namely, giving the emperor a diversified land base (*goryōchi*) – was not working out. So strategy was modified, and by the late eighties leaders were adopting a policy whose landed aspect focused almost solely on woodland. This they proceeded to assemble and organize with great vigor.

THE EARLY 1880S: A CONTEXT

As noted in Chapter 1, the 1870s rhetoric about freedom and people's rights (*jiyū minken*) was a surface expression of social turmoil so intense that it gave Meiji leaders ample reason to fear for the future of their regime and, therefore, to think creatively about measures to shield the Imperial Household, the instrument of their legitimacy, from that turmoil. Their problem had two major dimensions, fiscal and political, and both spurred efforts to secure the Household. Fortunately, the particulars of this political history have been well treated in numerous texts and monographs, so the matter need detain us here only briefly.[1]

Suffice it to say that by the late 1870s the fiscal demands of the decade – the costs of pensioning off declassed samurai, suppressing samurai rebellion, creating and funding new political and military institutions, establishing model industries, promoting new schools, and fostering new forms of learning – these several demands had thoroughly outpaced the government's tax-raising capability. Then the problem of excessive expenditures was exacerbated after 1877 by a decrease in income. That year, leaders slashed by some 16% the property tax rate they had established less than four years earlier, doing so in hopes of reducing popular anger and resistance to their reform policies. That move may have eased the discontent, but it exacerbated the fiscal shortfall and accelerated an inflation that was proving ruinous to the regime.

Unsurprisingly, perhaps, the deterioration in government finances was accompanied by escalating factionalism, tension, and violence in political life. And those trends clearly threatened the Meiji leaders themselves. Those leaders, as noted earlier, were "a motley crew," although their regime is oftimes characterized as a "Sat-Chō clique" because its leadership was predominantly samurai from the domains of Satsuma and Chōshū.[2] During the middle 1870s the key figures included Kido Kōin (Takayoshi) of Chōshū, Ōkubo Toshimichi of Satsuma, Ōkuma Shigenobu of Hizen, and the court noble Iwakura Tomomi. Kido's death in 1877 and Ōkubo's murder in 1878 left Ōkuma and Iwakura the two pre-eminent figures, soon assisted by Itō Hirobumi and Yamagata Aritomo, both of Chōshū.

Even before Kido died, the escalating political unrest had prompted these leaders to begin investigating mechanisms that would neutralize political activists by drawing them into constructive participation in the government through some sort of European-type assembly system. But the very question of how best to cope with those activists provided one more basis for factional conflict, and in 1880–81 it pitted Ōkuma against Iwakura, Itō, and Yamagata. In March 1881, he memorialized the emperor, proposing a relatively radical policy of reform that included the prompt drafting of a Constitution and creation of a parliament and cabinet of the British sort, with elections to be held within two years.[3]

Ōkuma's proposal precipitated heated debate among the leaders, but it also spurred them to try to reduce the public dissatisfaction with their policy. Most germane to our purpose was their decision to slow sharply the programs of *kan/min* division and woodland tax reform. They did so by the simple expedient of dissolving the office of tax reform in June 1881, declaring its task completed, and assigning the myriad unresolved issues to other government offices, thereby allowing the problems to be handled locally or to lie nearly undisturbed for the time being.

Meanwhile, Ōkuma's proposal, which most of his colleagues found far too radical, was producing such tension that by year's end he and his supporters had left the government. And soon thereafter they launched their own political party to agitate for constitutional reform. Other groups were similarly active by then, and they drew support not merely from the residue of disenfranchised samurai but also from urban entrepreneurs and more well-to-do villagers. People in those latter two categories not only had grievances against the regime but also aspirations to seize the moment and through political connections to advance their interests as captains of industry and leaders of hinterland reform and development.

As of late 1881, then, Meiji leaders – notably Iwakura and three Sat-Chō men, Yamagata, Itō, and Matsukata Masayoshi – faced two crucial dangers. Their regime was in dire fiscal straits and their leadership was being challenged by a burgeoning, radicalizing opposition movement.

They responded to the latter threat in several ways, most immediately by persuading the emperor to issue, on October 12, a rescript calling for the formation of a national assembly in 1890.[4] That deadline gave them a decade in which to study European practices, develop an overall reform strategy, devise means to contain more radical initiatives, draft a constitution, and make local and prefectural changes that would lay a foundation for orderly national elections to determine the assembly's composition.[5]

In the meantime they turned to the more pressing, if less exotic, task of repairing the inflation-ravaged fisc. Three days before the emperor issued his rescript, Matsukata had assumed the post of Finance Minister, and he shortly initiated a rigorous policy of retrenchment. To reduce spending, he scaled back or cancelled numerous government projects, including diverse programs of assistance to municipalities. He sold various government assets, reorganized banking to achieve tighter control of the money supply, sharply intensified an existing program to take money out of circulation, and hiked select commodity taxes.

During the next few years, these policies trimmed government expenditures appreciably while producing a sharp deflation that bolstered the real value of its tax income. By 1886, the fisc had been salvaged. However, the bulk of the price had been paid by poorer sectors of the hinterland populace. These were villagers whose agricultural output was selling at market for less than in recent years even though their taxes and rents were largely fixed, or in some areas even higher because of new levies that local governments had recently imposed to replace the assistance that Tokyo had ceased providing.

One unsurprising result of this drastic shift in fiscal policy and popular condition was an eruption of public protest that took the form primarily of peasant rebellions but also of broader participation in the freedom and people's rights movement.[6] The shift also exacerbated deforestation and downstream damage as government logged off areas hurriedly to generate income and villagers harvested sites recklessly for the same reason or to get out the goods before losing them through land-tax reform and *kan/min* division.

It was within this context of widespread social hardship, environmental injury, and political tumult that Meiji leaders addressed the task of securing their Imperial Household. The strategy they settled upon reflected this context, but it also reflected the new understanding of forests, forestry, and government's role in forest matters that was developing among them as a result of their contacts with Europe.

THE RATIONALE FOR GORYŌRIN

On the face of it, Meiji reformers had no good reason to regard woodland as a sound foundation for the Imperial Household. It was clear to anyone who traveled the realm that much of it was poorly stocked and minimally productive. That it was also enmeshed in a complex set of customary use rights was evident to any who made even modest inquiry. And early Meiji leaders surely were aware of a lesson long since learned by generations of lumber merchants: that theirs was an erratic and unreliable business. Burn down a city and business booms – if you have access to saleable timber. But once things are rebuilt, demand dries up, unused wood clogs the marketplace, timber providers have no way to move their goods, and bankruptcy looms.

Discovering the Forest Solution

What made forest look like a sound foundation despite these considerations were a new perspective and new knowledge that the reformers acquired from Europe beginning in the early 1870s.

Most notably, in 1871, Iwakura Tomomi led a large delegation, which included Kido and Ōkubo, to America and Europe to seek renegotiation of the existing commercial treaties. And much more broadly, they went to study those societies in hopes of learning the secrets of their wealth and power. Military, industrial, and political organization and technology were the core issues of interest, so one topic they investigated was the fiscal systems of European regimes. And the property arrangements of English and German royal families, with their grand estate houses, pastures, arable, and woodland, were one prominent aspect of that topic. A related topic that eventually caught their attention, at least fleetingly, was the condition and function of European forests and forestry more broadly.

For Kido and Ōkubo, silvicultural discovery seems to have occurred in Berlin. While there, Kido had to address the question of whether, given his government's tight budget, it should continue funding the Japanese students then studying in Europe. There was a large number of them, about half studying in Germany, but only one, Matsuno Hazama (Kan) was focusing on forestry.[7] In Berlin, Kido met his fellow Chōshū man Aoki Shūzō, a diplomat in the Japanese legation there, and when he heard from Aoki that Matsuno was studying forestry, he expressed skepticism about its usefulness, and therefore the wisdom of continuing to fund him. Aoki defended Matsuno's project as of value to government and people alike, and he had Matsuno come and explain to Kido and Ōkubo the logic of his choice. Evidently the student was persuasive – surely the fact that he too was from Chōshū helped – because his funding was continued until he completed his program in 1875.

Matsuno's case may have been strengthened by the fact that he was getting the best training available. At the time, forestry experts in Germany, together with those at the French school of forestry in Nancy, were at the cutting edge of research and practice in the lively new discipline of regenerative forestry. And Matsuno came to share their sense of professional enthusiasm and confidence. Doubtless, too, the appeal of this German forestry was enhanced by its cachet of up-to-dateness. Even in England and British India, German silviculture enjoyed considerable prestige during the later nineteenth century as the best path to revitalized timberland.[8]

Furthermore, the German school emphasized the importance of sustaining timber production over the long term, which in their view meant giving government a pre-eminent role in woodland planning and control. It was a viewpoint that fitted nicely the "developmentalist" perspective that Ōkubo and other mission members were coming to share.[9] After Matsuno returned to Japan in the autumn of 1875, Kido and Ōkubo drew him into the newly developing government forest agency, where he went on to a distinguished career as a key architect of Japan's modern government-steered forestry.

This early 1870s exposure to European silviculture offered these men a number of insights that pertain to our inquiry here. They include a recognition of the importance of timberland to industrial society, a new awareness of Japan's extraordinary wealth of woodland acreage, and a realization that there were means of "scientific" management by which foresters could repair deteriorated forest and enhance its stability and long-term productivity. As a consequence Japan's woodland could, if properly managed by trained professionals, produce the larger timbers and other goods required by a modern industrial society and thereby serve as a stable, durable, slowly-growing source of wealth for the Imperial Household.

Japan's own well-developed tradition of forest thought and practice was, in fact, quite nicely adapted to the archipelago's woodland biome. But old wisdom such as that was generally being repudiated by the "progressive" voices of early Meiji. The general experience of Iwakura Mission personnel reinforced that attitude. It helped to give proponents of the new forestry a firm sense that theirs was the silvicultural wisdom appropriate to the times and essential to the future and therefore something properly promoted regardless of objections by others. It was a wisdom, after all, that would enable foresters to repair the realm even as they secured the Imperial Household: what was good for the emperor would, indeed, be good for the people.

Advocating the Forest Solution

Senior members of the Iwakura Mission returned to Japan during 1873. Ōkubo then became a particularly strong advocate of a vigorous system of government forest (*kokuyūrin*). Because he headed both the Interior Ministry and the land-tax reform program, his priorities shaped policy. Both he and Kido also began to promote woodland's role as a source of income for the Imperial Household more specifically, with Kido explicitly urging in 1876 that government forest be used for that purpose.[10]

During the next few years, other figures of similar stature supported that view, some urging that all government forest be so used. It was a posture strengthened by the spread of modern forestry concepts and research activity within Japan.

Particularly salient was the German concept of "Normal Wald" (*hōseirin*), the ideal of a forest so managed as to assure a sustained yield year after year. Matsuno and others were promoting the concept as a basic guideline for national forest policy.[11] If applied skillfully in programs of afforestation, aftercare, and regulated felling, they argued, that concept would assure a perpetually regular harvest and thus a reasonably regular income. Or at least a reliable annual increment of standing timber and a regular annual cut that could be milled and stored for eventual sale under optimal circumstances.

During the eighties the conviction that progressive "scientific" policy could make woodland a source of expanding and dependable income rapidly gained favor. Surely the idea found such ready acceptance because many particulars of silvicultural technique, as well as the basic notion of purposefully managed forest, were similar to practices and viewpoints already found in Japan's indigenous silvicultural tradition. But now they were rationalized and codified as up-to-date policy and practice in the new and growing field of modern forestry.

Earlier, in the spring of 1878, Matsuno and others had established a forestry research station at Nishigahara on the northern outskirts of Tokyo; in following years its operations grew. Opened originally under Interior Ministry auspices, it was reorganized in December 1882 to include instruction as well as research. Then, with the title Tokyo School of Forestry, it was made a unit of Tokyo Imperial University. As the eighties advanced, the School launched more and more forestry research projects and produced a growing number of experts for government employment. Meanwhile a proliferating array of local and prefectural forestry associations (*sanrinkai*), with encouragement from the National Forestry Association (Dainihon Sanrinkai, founded in January 1882), were undertaking their own projects and disseminating the new silvicultural understanding widely throughout the realm.[12]

The growing conviction that forest could provide a reliable and gradually expanding harvest was seized upon by those advocating a securely-funded Imperial Household. Most notably, in May 1880, a year before his final falling out with other leaders, Ōkuma drafted a wide-ranging reform proposal.[13] In his third point, he called for establishment of a large imperial estate. He advised that it be composed of forest (*sanrin*) because, "that will be easy to administer. Moreover, by planting and nurturing the forests, they will improve naturally."[14] He highlighted the magnitude of Japan's woodland by noting that, "government forest covers about 7,267,383 *chō*, or some 4,612 square *ri*" (more than 70,000 km^2). He then cited production and income figures for 1877 to show its profitability, continuing in this way:

> The Forest Agency [Sanrinkyoku] estimates that in 1882 the balance [in its operating budget] will be about 52,183 yen, with a savings on timber of 393,943 yen or a total profit of 446,126 yen. Thereafter, depending on managerial technique, the level of income and expenses will rise or fall. But as transportation improves and if planting is well handled, income should not vary greatly from these estimates.

Ōkuma had gone to considerable length to demonstrate the magnitude of *kokuyūrin's* potential as the base for an imperial fisc. And in asserting that it would be easy to administer and a source of substantial and expandable wealth, he was making points that proponents of Goryōrin would emphasize thereafter.

Developments of the next few years intensified the discussion and spurred decision-making. Specifically, as noted above, public unrest was in the process of forcing leaders to develop electoral institutions in the hope that those would blunt the discontent and strengthen public support for the regime. Before allowing electoral politics to shape government policy, however, those leaders wished to secure the Imperial Household's independent fiscal foundation.

In that spirit, late in 1881, even as factional bitterness was flaring, Iwakura and Inoue Kaoru, a man from Chōshū who had long been involved in Meiji fiscal matters, reiterated in separate proposals Ōkuma's suggestion that all *kokuyūrin* be assigned to the emperor. And, in January 1882, a senior official in the recently constituted Ministry of Agriculture and Commerce (Nōshōmushō) prepared an elaborate plan for assigning forests, other land, mines, docks, and factories to the Imperial Household.

In February, just before leaving for a year in Europe, Itō proposed that the realm be conceptualized in tripartite terms as property of the emperor, the government, and the people. Iwakura added his voice again, recommending that all government woodland, plus all the government's railroads and factories, be assigned to the Household.[15] Others presented similarly expansive ideas, but *kokuyūrin* continued to be seen as an important part of the scheme. It was the property that would be most durable and easiest to manage.

In these proposals some commentators looked especially at the recently acquired, nearly pristine reaches of Hokkaido as an ideal realm for Goryōrin. But a more ideology-oriented perspective also emerged, one that called for creation of an imperial domain centered on Mt. Fuji, with its resident deities and sacred aura. Surely the appeal of this latter proposal was enhanced by its geographical practicality. It would situate Goryōrin directly between the major timber markets of Tokyo and Nagoya, in a region favorable to the most highly prized species of tree, notably cryptomeria (*sugi*) and Japanese cypress (*hinoki*). In any case, these lines of thought – assemble a mixed portfolio of assets, give it ample woodland, including the vacant realm of Hokkaido, and center it on the great symbol of Mt. Fuji – were key components of the initial scheme for a large *goryōchi*.

The linkage between Japan's nascent discipline of silvicultural research and the political agenda of a securely funded Imperial Household was most fully expressed in the person of Nakamura Yaroku. He was a professor in the new Tokyo School of Forestry and also an advisor to the Imperial Household Ministry, as well as a member of the delegation that accompanied Itō to Europe in 1882. After a year abroad

Fig. 2-1. Kiso *kanrin.* This plantation stand of *sugi,* adjacent to a river in the Kiso Valley and photographed during early Meiji, exemplifies the type of forest that creators of Goryōrin envisaged as the foundation of their Imperial House. Note the steepness of hillside visible through the trees, lower left. (Reproduced courtesy of Dean John C. Gordon, from a set of photographs in the Yale School of Forestry and Environmental Studies.)

Fig. 2-2. Desirable *sanrin*. Although this particular site (near Ōme, west of Tokyo) was never a portion of Goryōrin, the scale, topography, and stand composition exemplify the *sanrin* that builders of the Imperial Forest preferred to include. (From author's collection.)

studying political institutions and other matters, Itō's group returned to Tokyo in August 1883, with Itō and Nakamura firmly convinced that forest would make better *goryōchi* than would other properties.[16] Nakamura promoted that view vigorously thereafter, and one of his proposals quickly became a policy guideline.

Specifically, Nakamura advised that in its landed aspect the emperor's fiscal base should consist essentially of woodland and that it should contain only large parcels, 100 *chōbu* to be the minimal size. The 100-*chōbu* (99 hectare) size was a reasoned choice. It would assure foresters parcels sufficiently large to grow tall, straight trees rather than the stubby, scraggly "wolf trees" that tend to dominate small parcels exposed to lateral sunlight. It would permit the establishment of multi-segment, rotation-cutting programs, and would generate harvests large enough so that logging operations could achieve economies of scale in both their thinning and final cuts.

The notion that forest would make particularly valuable imperial estate continued to be advanced during the later eighties.[17] And indeed, to get a bit ahead of our story, the soundness of the woodland choice was validated during the nineties by the fact that even as other properties continued losing money, forested acreage began generating a tidy and growing profit, an outcome that reflected attentive management as well as burgeoning demand for construction timber.[18]

FROM GORYŌCHI TO GORYŌRIN

Starting in the early eighties these calls to give the Imperial Household some sort of landed base were acted upon. From the outset, imperial estate lands were viewed as a fiscal foundation that would help to insulate the Imperial Household from the vagaries of popular whim and political fashion. But how extensive that estate should be, what form it should take, how it was to be administered, and what portion of the Household's entire foundation it would be – these were all issues to be sorted out.

The Expansion of Goryōchi

As the eighties advanced, more and more productive acreage was assigned to the Imperial Household. By 1885 its 634 *chōbu* of 1880 had grown to some 22,000 *chōbu,* a figure that continued rising thereafter, approaching 40,000 *chōbu* before the decade's end.

All the rhetoric about woodland notwithstanding, however, prior to 1889 very little well-stocked forest was actually assigned to the Household. Probably this was so because little woodland had been adequately surveyed due to the widespread disputes over boundaries and use rights that had temporarily stalled the policy of *kan/min* division back in 1881. In consequence, the best combination of sites for inclusion in an imperial estate could not be determined. It also seems likely that Meiji leaders did not want to risk entangling the Household in those disputes. So the land that the Household received came to it in a helter-skelter manner. Most of it was arable and pasturage, but it also included scattered fisheries, bits of woodland, and eventually mines and even a few industrial plants.[19]

It gradually became evident, however, that the imperial estate as then constituted was failing in its intended fiscal purpose. Except for a cattle-rearing enterprise in Hokkaido, all of the properties turned out, for diverse reasons, to be more or less unprofitable. Consequently, from the later eighties onward, Household officials transferred more and more parcels of their arable, pasture, and other land back to prefectural authorities, who commonly leased them out to cultivators or other users as rental property. Even the fabled bullion mines of Sado and Ikuno, which the Household Ministry received in 1889, along with the enterprises acquired or built to go with them, proved disappointing and were sold in the next few years.

By 1892, Ministry officials were so disillusioned with their multi-asset base that when they drafted surveying guidelines for use in identifying parcels to be permanently retained, they formalized their preference for forests by distinguishing clearly between woodland and other properties. The former were to be retained under direct supervision while all others, whether used for tillage, grazing, mining, salt-making, or fisheries work, were all to be leased.[20] And in later decades, as we note in Chapter 3, nearly all those leased parcels were sold or bequeathed to prefectural authorities.

Shifting toward a Goryōrin Policy

By the early 1890s, then, the goal of a multi-asset estate had been thoroughly repudiated. The actual shift to a woodland estate had begun a few years earlier, as the shortcomings of non-forest parcels started becoming apparent.

In March 1884, Itō had assumed headship of both the Imperial Household Ministry and a committee charged with reorganizing government prior to the implementation of parliamentary elections. Those appointments placed him in a strong position to shape policy, and so to advance the woodland ideas he and Nakamura had acquired in Europe.

Late in November 1885, Itō, acting in his dual capacity as Household Minister and Head of Government Reorganization, moved to preserve desirable forest sites by calling on the Ministry of Agriculture and Commerce to halt all further sales of *kokuyūrin* parcels exceeding 100 *chōbu*. Such sales had been occurring occasionally, especially in Tōhoku, as part of the tortuous process by which woodland was slowly surveyed, boundaries delineated, local grievances meliorated, and any unpromising parcels that fell within *kokuyūrin* borders deemed expendable and sold for income.

Itō's request was heeded, and during the next few years, as officials of the Forest Agency (a unit of the Ministry of Agriculture and Commerce) pressed ahead with their surveying of *kokuyūrin*, they focused on areas that seemed likely to be assigned to the Household Ministry. In this work, and in the later assigning of woodland to the Ministry, the 100-*chōbu* minimum remained a guideline that helped shape decisions.

By 1885, questions had also arisen about how to administer the growing imperial estate. That year, leaders agreed that an independent office within the Household Ministry, rather than a regular government office, should handle the task. Accordingly, late in December – just a day after Itō took fuller control of the government with the newly-created position of Prime Minister, which he held concurrently with his existing title of Household Minister – the government established a Property Office (Goryōkyoku) within the Household Ministry for the purposes of administering its assets and handling its financial affairs.

Initially the new office consisted of two departments, one for accounting and one for everything else. But a few weeks later a department of land survey was added, and shortly afterward three regional headquarters were established to oversee the household's scattered parcels. By the end of 1886, an administrative structure for handling additional land was being developed, and in following years it expanded. By 1889, the Property Office had a permanent staff of thirty-three assisted by a variable number of temporary subordinates.[21]

In sum, what had begun in the later seventies as an ill-focused intent to forestall political danger by securing an independent fiscal basis for the Imperial Household had, by 1890, sharpened, in its landed aspect, into a resolve to create a mighty Imperial Forest, Goryōrin. In pursuing that objective, Meiji leaders had, in effect, displaced the European-inspired dyadic construct of "public" and "private" property, which had initially informed their whole venture in land-tax reform and *kan/min* division, with the triadic construct of imperial, government, and people's property that Itō had proposed in 1882.

ASSEMBLING GORYŌRIN

One suspects that the creators of Goryōrin viewed their task as fairly straightforward. Once the process of *kan/min* division was well enough advanced, the Forest Agency and prefectural surveyors would utilize its results to identify parcels of *kokuyūrin* worthy of transfer to the Imperial Household. These would be reassigned to the Property Office, and the organized whole would then be managed in a manner that assured its long-term profitability for the Imperial Household. Whatever the expectation, however, getting to that outcome proved a very vexing task.

The Process of Assemblage

From the outset, the process of assemblage encountered difficulties, with government agencies, both central and prefectural, finding themselves at odds, sometimes harshly, about the desirability of using particular parcels for Goryōrin rather than other purposes. One reason for disagreement was that transfers of woodland to Goryōrin status deprived prefectural authorities of future income; another was that they complicated Forest Agency efforts to administer nearby government forest. The difficulties notwithstanding, however, authorities pressed ahead, and two years of labor seemed to give them the woodland estate they sought.

The actual process of selection commenced in May 1888 when a coordinating committee drawn from the three relevant ministries (Agriculture and Commerce, Interior, and Household) began meeting to identify sites for inclusion.[22] They focused their attention on the best portions of *kokuyūrin*, most of which the Forest Agency already controlled directly as *kanrin*. Unsurprisingly, the most celebrated forests of the Tokugawa period – in present-day Nagano, Shizuoka, Akita, and Aomori prefectures – were on their list, along with woodland in several other prefectures and the Kyoto-*fu* hinterland. The areas they initially proposed were registered as 2,270,000 *chōbu*, nearly one-tenth of Japan's woodland.

At once, the proposal encountered opposition from Forest Agency officials administering the large areas of *kanrin* in Akita and Aomori. They warned that reassigning select portions would confound their ability to manage what remained. So the conferees dropped all of Aomori, nearly all of Akita, and several other contested sites. They located some substitute areas and by that October had agreed on an assemblage of parcels scattered among thirteen jurisdictions from Sendai to Kumamoto. But they totaled only 908,108 *chōbu*.

In both scale and coherence that committee's final effort was unsatisfactory to Itō and others. So in following months a larger and more high-powered inter-ministry group, with Nakamura included, examined the issue to assess its tax implications. In its deliberations this group proposed that Goryōrin be clearly centered on Mt. Fuji, that peripheral areas be deleted, and that more acreage be added, especially in the Kiso, Tenryū, Ōi, Fuji, and Sagami river valleys (see Map 4). Changes were made; more than 400,000 *chōbu* in Yamanashi were added, and by September of 1889 a decision had been reached that identified 1,427,797 *chōbu* of assignable forest, most of it centered on Fuji. By year's end the parcels had all been formally transferred to the Property Office, most of them *kanrin* that had been under Forest Agency jurisdiction.

Meanwhile, Itō continued to press for more acquisitions. And the Household Ministry gathered information directly from prefectural officials regarding potential sites in the Kantō and Tōhoku districts and subsequently in central and western Japan. During 1889–90, after considerable backing and filling, these initiatives yielded another 153,019 *chōbu* scattered widely about the realm. Most of the parcels were forest of no special note, and most were in Tōhoku. A few, however, were sites rich in sacral lore, including the small but storied woodlands of the imperial family's Ise Shrine in Mie Prefecture and the forests of Mt. Takachiho in southern Miyazaki. The latter was celebrated as the place where Ninigi no Mikoto, mythical founder of the reigning dynasty, descended to earth from the Plain of High Heaven.

Two years of work had yielded much, but the Household Ministry wished more. During the 1880s, it had accumulated some 31,350 *chōbu* of pasturage and arable in Hokkaido, and at its request, in 1890 an area thought to encompass about half of Hokkaido's *kokuyūrin*, and arbitrarily listed as 2,000,000 *chōbu*, was added to that sum and assigned to a newly-opened regional headquarters in Sapporo. With those transfers the Imperial Household had acquired a putative total in excess of 3,600,000 *chōbu* of land, nearly all of it forest of varying quality. As Itō wished, the realm had indeed acquired something of a tripartite character of imperial, government, and people's lands.[23]

Quantitatively the effort had yielded an impressive woodland estate. Qualitatively, however, all the attentiveness notwithstanding, the outcome was less gratifying. As Household Officials were quick to discover, their properties were ill-identified and of highly inconsistent character and value. Indeed, they were so varied that they continued to be referred to not as *goryōrin* (imperial forest) but by the more general term *goryōchi* (imperial land) or, very commonly, *goryō rin'ya* (imperial forest and field), where the *ya* connoted not only open terrain but also relatively flat land in contrast to the sense of woodland as essentially steep hillside (as in *sanrin*: "mountain-forest").

The principal chronicler of Goryōrin has summed up the situation concisely, writing that the original accumulation included:

> Extremely small parcels, common-use land (*iriaichi*), areas with shared stands (*bubungi*), sites with production rights already conceded, leased-out

> parcels, and areas in the process of being petitioned for disposal (*sagemodoshi*) as people's land.[24]

Indeed, many of the parcels were already the focus of intense litigation between local users and government, most widely over issues of common-use rights. In consequence, as Property Office personnel labored to organize and establish orderly management practices in their realm, they found themselves enmeshed from the outset in seemingly endless policy clarification, negotiation, lawsuits, petitions and their appeal. And some of the disputes dragged on into the 1920s.

Although woodland proved in the end more manageable and more profitable than a multi-asset base, Goryōrin thus began as a messy assemblage of variegated, conflict-ridden land, especially in Naichi ("the inner regions," here meaning Japan south of Hokkaido).[25] In consequence, even as the Property Office acquired jurisdiction over vast, new territories during 1889–90, it found itself drawn into intensified programs of land surveying and further organizing of its central and regional administrative units.

Organizing the Assemblage

Acquiring land was one thing; managing it, quite another. The process of acquisition had given the Property Office far more acreage than it could actually handle, so it focused on solidifying control of the most desirable parts, leaving the rest under supervision of regular prefectural authorities.

During 1889, even as land selection advanced, the legal status of Imperial Forest was clarified. On February 11, the day the emperor formally announced the new Meiji Constitution, the Imperial Household Ministry enacted its own "private" House Law to regulate all matters of the imperial family and its estate. Regarding land, it specified that "Imperial writ" would determine which of its properties were to be designated "hereditary" (*seden goryō*) and declared that:

> No landed or other property, that has been fixed as the Imperial Hereditary Estates, shall be divided up and alienated.[26]

Other, less valuable parcels of *goryō* would be held in a more negotiable status until they were adequately surveyed and evaluated and decisions made about their suitability for permanent retention or disposal.

With elections for a national Diet in the offing, however, Household officials wished to get as much land as possible designated *seden goryō* before Diet members could meddle in the process. Accordingly, they made hurried attempts to survey their holdings during 1889–90. And in November 1890, just before the Diet convened, the Ministry declared areas registered at 1,010,188 *chōbu* to be hereditary *goryō*. The properties were scattered over twenty-eight locations, mostly in the Kiso River valley and flanking Mt. Fuji in Yamanashi and Shizuoka prefectures. None was in Hokkaido, which still awaited surveying. By 1891, then, to considerable degree the Imperial Household had its Fuji-centered fiscal base safely secured against political manipulation.

The Property Office also moved expeditiously to create and staff regional headquarters for managing its best forests: one in Kiso, another in Shizuoka to handle its Fuji-centered *seden goryō*, and a third in Sapporo to oversee its vast Hokkaido dominion. Until the late 1890s, however, most of its properties in Yamanashi Prefecture, the Kantō-Tōhoku region, and other scattered locations remained under varying degrees of prefectural control for want of regional organization sufficient to supervise them.

Meanwhile, officials at the regional headquarters and in prefectural offices pressed ahead with the task of surveying their respective parts of Goryōrin – determining boundaries, erecting markers, estimating acreage, and evaluating stand volume and quality. Much of the mensuration was crude, especially in Hokkaido. But in Naichi the tasks were largely completed by 1896, although some areas took another decade to finish, mainly because of legal disputes. In Hokkaido most of the basic surveying was done by 1914.

By then, however, as we see in Chapter 3, that very surveying activity had contributed to a broader trend that reduced the size of Goryōrin, even as it organized and consolidated it and helped steer its yield into imperial pockets.

3

Reducing Goryōrin

Almost from the outset the Imperial Household's new forest arrangements ran into trouble, and those difficulties eased only slowly as decades passed. In many instances the Property Office dealt with its problems – whether willingly or due to external pressure – by disposing of the troublesome parcels of land.

To adumbrate the cumulative statistical impact of that strategy, by 1897 it had reduced Goryōrin acreage from a claimed 3.6 million to 2,228,890 *chōbu*. And the figure kept dropping thereafter: to 1,933,574 by 1918, and 1,349,505 in 1946.[1] For the record, the 1937 acreage of 1,413,303 *chōbu* was scattered about the country as indicated in Table E-2 (in Appendix E).

By the 1930s, the Imperial Forest was thus a much smaller institution than its founders appear to have envisaged. But the process of shrinkage had eased many grievances. And it had been handled in a manner that left the Household with a valuable and highly profitable estate, as we shall see in later chapters.

THE OVERALL PICTURE

Several factors contributed to this long-term shrinkage, and both officialdom and villagers played a role in it. Only a few people opposed Imperial Forest in principle; most objected for practical reasons to Household control of specific parcels or to regulations and practices that threatened their interests in or near such sites. Resistance came from elected and administrative officials in Tokyo, from officials in prefectural and district offices, from entrepreneurial lumbermen, and, most persistently, from villagers all across the realm. Officialdom resisted through administrative action and political maneuver; others, through petition, protest, sabotage, and hostile alliance.

Initially, losses of Goryōrin resulted primarily from external pressure, but from about 1900 onward, more and more were outcomes of administrative choices made by officials within the Property Office. In some cases they disposed of property to reduce local political difficulties. But more frequently as the years passed, they did so to improve the quality of their forest operation by selling unwanted parcels as a way to finance improvement work on what remained.

Of course, part of the shrinkage in Goryōrin acreage was merely a statistical artifact produced by ever more careful land surveys. But a greater part was real losses that reflected widespread resistance.

What portion of the total was a bookkeeping artifact is unclear. As shown by the Hokkaido case that we note below, after all, resurveying could disclose major underestimates as well as overestimates. In any case, careful boundary surveys did reveal gross errors in the original accounting, most commonly overestimates in Naichi. That outcome reflected four basic facts: most woodland had never been carefully surveyed; the surveys of 1889–90 had been made hurriedly and relied primarily on written records rather than actual measurement; properly trained surveyors were in short supply; and accurate surveys of tortuous mountain terrain were extremely difficult to conduct even with the best equipment of the day.

A major part of the shrinkage, however, was due to real losses. Basically, one might argue, they occurred because proponents of Goryōrin were at a severe disadvantage, their high status notwithstanding. They were few and far away, being little more than senior figures in the government and the staff of the Household Ministry and its subordinate units. Those opposed to transferring woodland to the Property Office, on the other hand, were legion. Many were well positioned to resist, and many continued laboring to undo such transfers long after they had been made.

The single largest loss was also one of the earliest, and it illustrates how the fate of Goryōrin could become hostage to other issues of the day. The case involved Hokkaido, where, in April 1893, the governor complained to Tokyo that the recent transfer of land to Goryōrin status was obstructing his program for opening the island to settlement, cultivation, and other uses. He recommended that the Hokkaido areas be exchanged for *kanrin* in Naichi. His message elicited action because it spoke directly to one of the highest-priority political concerns of the moment: relations with Imperial Russia.

To explain, Meiji leaders had inherited from the Tokugawa a strong fear of Russian expansion toward Japan, and ever since the Restoration speedy settlement and development of Hokkaido as a buffer zone had been a top-priority policy. By the early nineties, Japan's leaders were embroiled in high-stakes maneuvering for geopolitical advantage in the vicinity of Korea, and they regarded secure control of Hokkaido as essential to success in this larger enterprise. The warning that Goryōrin on the island hampered the project led to a prompt reconsideration. In October 1894, even as Japanese troops were battling Chinese for dominance in Korea, the Household Ministry surrendered 1,370,000 *chōbu,* keeping only an estimated 630,000, even though no replacement acreage was found in Naichi. Later resurveying revealed that Hokkaido woodland as a whole covered some six million *chōbu,* not the four million originally postulated, and that the residual Hokkaido Goryōrin actually held some 900,000. So it seems likely that the original assignment was nearer three than two million and the real loss in 1894 much more than the official 1.37 million.

Changes elsewhere were less grand in scale, but they proved more disruptive in the longer run. The prized *seden goryō,* which initially had been recorded at 1.01 million *chōbu,* was within a few years restated at only 401,712. And, by the 1930s, it was a shrunken remnant of the original Fuji-centered vision. In 1937 it measured only 208,478 *chōbu,* mostly in old Tokugawa woodlands of Nagano, Gifu, and Shizuoka prefectures, and it was centered on the Kiso River.[2]

Non-hereditary *goryōchi* also fared poorly, though less so. Its greatest losses also were in the prefectures near Mt. Fuji. In the outcome, non-hereditary acreage stabilized at about 1,200,000 *chōbu,* of which 900,000 were in Hokkaido, where Imperial Forest never received *seden goryō* status. Nearly 150,000 *chōbu* lay on Honshu north of Tokyo, and less than 120,000 were near Fuji (in Nagano, Shizuoka, and Kanagawa prefectures). The remainder lay scattered widely about the realm. By the 1930s, nearly two thirds of the total Goryōrin acreage lay in Hokkaido, with the remaining one third spread across Honshu from Gifu Prefecture north-eastward to the far corners of Aomori.

RESISTANCE AND ITS EFFECTS

The magnitude of shrinkage in Goryōrin was a fair measure of the resistance to it. People had many reasons to oppose transfers of land to the Property Office, and often they collaborated in their resistance. The character and role of key protagonists may come clearest, however, if we examine the opposition in terms of officialdom and villagers.

Resistance from Officialdom

Both officials in the central government and those in the hinterland, at prefectural, district, and local levels, raised objections to Goryōrin. And their actions clearly shaped the outcome of the Imperial Household's great enterprise. The Hokkaido case exemplified, in extreme form, how the interplay of central and regional interests could influence policy on Imperial Forest. But central and hinterland influences could also be effective on their own.

1. *Central Officialdom*

Both during and after the initial creation of Goryōrin, administrators in key ministries, those of the Forest Agency most importantly, were able to shape decisions on its composition and handling. And, after 1891, both the Supreme Court and Imperial Diet weighed in, often to the discomfort of Household officials.

The Supreme Court was repeatedly asked to resolve questions of law regarding claims on government land, and during the nineties, as we note briefly below, it issued some judgements that affected Household interests. Elected Diet members, whose constituency politics prompted resistance, occasionally challenged policies of both the Forest Agency and Property Office, and one of the most energetic participants in Diet debates about woodland was the redoubtable Nakamura Yaroku. Having left the government and been elected to a seat representing richly-forested Nagano Prefecture, he used his considerable stature and expertise to help fashion legislation that was conducive to sound forest practice while also being responsive to the lumbermen and landowners whom he represented.[3]

Central officialdom's discontent with Goryōrin was most evident during the nineties. It was never fully stilled, however, and reappeared later when fears for the imperial institution weighed less heavily than frustration with the fractured forest system. From about 1917 onward, even as Goryōrin continued to shrink, occasional petitions from professional foresters and others requested that it and *kokuyūrin* be recombined. In 1928, during his rituals of enthronement, the incoming Shōwa Emperor himself asked an Imperial Household official about the possibility of placing Goryōrin under Forest Agency supervision. Household officials responded by explaining the reasoning that had originally led to Goryōrin's creation, and the young emperor let the matter pass.[4]

The imperial query suggests, however, that even at the highest level the administrative illogic of Goryōrin was recognized, and as the 1930s advanced, some Forest Agency officials advocated more openly its reintegration with *kokuyūrin*. They reasoned that the two forms of woodland were administered in essentially the same way, so the redundancy should be eliminated, thereby improving efficiency and effectiveness of operation.

This trend of thought, if such it was, faded after 1937 when the war with China escalated. As that conflict evolved into the Pacific War, however, and as both military demand for timber and military control of the economy intensified, the Property Office and Forest Agency found their policies being subjected to more and more outside guidance, which further eroded the logic behind their formal separation.[5]

When the issue of forest governance was addressed again in 1946, after the war's end, the occupying Americans viewed the issue not in terms of administrative efficiency, but from the perspective that created Goryōrin in the first place, that of political ideology. The Imperial Household, they averred, ought NOT to have a fiscal base insulated from the electorate. So *kokuyūrin* and Goryōrin were finally reintegrated under Forest Agency control, as so many foresters had long desired. But it had been accomplished only thanks to defeat and foreign occupation, and it had been done as an explicit repudiation of the logic that had originally given rise to an imperial estate.

2. *Regional Officialdom*

Another important source of dissent during Goryōrin's early years was prefectural and district officials in Naichi, who often opposed woodland transfers. Their view counted because in many areas they retained actual control of *goryōchi* parcels for years after the land's formal transfer to Property Office jurisdiction. They commonly opposed transfers to Goryōrin status for much the same reasons they resisted assignment of *kan'yū rinya* to direct Forest Agency control as *kanrin*. In part, surely, it was generic regional resistance to centralization. But more substantively, regional officials wished woodland converted to tillage wherever feasible, or otherwise developed in ways that would benefit the prefectural or local fisc. Yielding parcels to the Household Ministry risked permanently losing their income potential.

In addition, regional officials in Naichi realized that local people would in most cases lose income, employment, and daily necessities if woodland were shifted from existing multiple-use arrangements into long-term reforestation. In consequence, when parcels were transferred to the Ministry despite their objections, prefectural leaders commonly used their political clout to preserve such local uses of the land as already existed. Doubtless they did so in part to protect their regional tax base. But in greater part, one suspects, they did it to minimize public discontent that could complicate their governance, jeopardize their careers, or offend their sense of purpose and propriety.

Resistance from Villagers

Most important in confounding Property Office policy in Naichi were villagers, who generally opposed the inclusion of any nearby woodland in Goryōrin. Often, their opposition did not emerge at once because they might not discover very quickly that nearby parcels had been transferred from *kokuyū* to *goryō* status. Nor did they necessarily grasp what such transfers could mean. The whole business, after all, was being handled by officials at the central and prefectural levels, who corresponded with officials at the district and administrative village level while having almost no contact with locals. Moreover, the fact that prefectural officials continued to handle much Goryōrin during the nineties helped conceal the status changes.

When villagers did become aware of such changes, however, they commonly had grounds for dismay and protest. The issue at stake was not some romantic attachment to ancestral lands or some aesthetic devotion to beautiful surroundings, but rather the fact that forests were crucial to their daily lives.

Although the hillsides that were being transferred were ones the government already claimed as *kokuyūrin*, in most locations they were still being used for village purposes. Throughout Naichi, as noted in Chapter 1, villagers had long enjoyed a wonderfully complicated set of commonland use rights on both woodland (*sanrin*) and waste (*gen'ya*). Sometimes hillsides were the sites of shared-yield timber stands (*buwakebayashi* or *bubungi*), but most commonly villagers regarded them as commonland (*iriaichi*) of one sort or another. And the sites provided them with a grand variety of necessities, most notably green fertilizer, fuel, fodder, pasturage, and small-scale construction timber.

Fig. 3-1. *Gen'ya*: contested land. The brushy border areas of this location (southwest of Tokyo) exemplify the innumerable patches of *gen'ya* that were contestable ground. Commonly identified as *iriaichi*, their potential for different uses pitted villagers against the creators of *kokuyūrin* and Goryōrin. As *gen'ya*, such sites yielded diverse goods, but for villagers they also had potential as swidden, pasture, or even dry fields. For the government, on the other hand, they had good potential as easily accessible timber forest. In terms of the *ringyōchi* categories discussed in Chapter 5, if this site had been Goryōrin, the arable area would have been classified as non-timberland and the brushy area as potential timberland. (From author's collection.)

During the 1870s–1880s they had insisted on the legitimacy of those rights even when lacking the documentation required by officials engaged in implementing the *kan/min* division and land surveys. Where the government had effectively acquiesced in their claims by writing old uses into new leasing arrangements, villagers retained those use rights on *kokuyūrin*. But commonly they would still protest the costs and restrictions that came with the new agreements. And in many cases where the government had denied their claims, villagers actively challenged the very designation *kokuyūrin* or resorted to one or another "use it, don't lose it" stratagem. The re-designation of such parcels as Goryōrin seemed further to endanger village interests, especially because nearby hillsides often were the very ones that could most profitably be converted to timber production and therefore seemed most attractive to the builders of Imperial Forest as future *seden goryō*.

The Effects of Resistance

Prefectural authorities commonly supported these local use-rights claims, and in the face of such combined resistance the Property Office tended to be conciliatory, especially in outlying areas such as Tōhoku. On many parcels it agreed to stand-sale and land-lease arrangements that constituted functional acceptance of existing practice in return for villagers' formal acquiescence in the Household Ministry's legal claim to jurisdiction.

The situation is exemplified by the regulations issued in 1889 by officials of Gunma Prefecture (see Fig. D-1 in Appendix D), who were handling woodland parcels newly designated as Goryōrin on behalf of the Property Office. While those regulations assumed an official right to administer the land, they also accommodated existing use rights through a vocabulary of land-rental and stand-purchase agreements. But they also reveal the tightening of control and the burdensome procedural requirements that so irked villagers.

The Property Office was not, however, equally concessive everywhere. It was most intractable, it appears, in handling areas deemed crucial to the overall goal of a Fuji-centered hereditary *goryō*. The most traumatic struggle of all, that over Yamanashi Prefecture, we examine in Chapter 4. Better known is the case of the Kiso River valley in southern Nagano Prefecture.

1. *The Case of Kiso*

In Kiso, villagers had from the outset been trying to recover use rights on *kokuyūrin* that they had lost during the 1870s *kan/min* division. The problems simmered during the eighties, but new land surveys of the nineties, which followed assignment of the area to Property Office jurisdiction, revived local dismay. In June 1898, the locals tried again, their hopes lifted by two legal developments. Property Office regulations of that January had seemed to facilitate the sale of unwanted Goryōrin, and a court ruling in May recognized *iriai* rights in *kokuyūrin*. The ruling carried no weight in Imperial Forest, but villagers may have not grasped the distinction, or perhaps they hoped to establish the converse.

Seizing their perceived opportunity, representatives of sixteen villages prepared a petition seeking the restoration of use rights on some 280,000 of the 350,000 *chōbu* of Kiso forest then registered as Goryōrin. District officials, who recognized the importance of those areas to villagers' livelihoods, assembled elaborately detailed data to support the petition, including the specifics of procedures by which villagers would protect the woodland from misuse. Prefectural officials then added their support, and villagers provided more documentation of their claim.

However, the areas being claimed (known as *akeyama* or "open forest") were universally recognized as having been lord's forest (*ohayashi*) of the Owari daimyo. And while it was also clear that the villagers had long enjoyed customary access to those *akeyama* areas for diverse purposes, in the end they were unable to provide documentation showing a long-standing practice of nurturing the woodland. Property Office rules stipulated that without evidence of a custodial investment in forest maintenance, villagers could not claim an accumulated right of exploitation. So in the autumn of 1899 the petition was denied. Nor did further efforts during the next two decades reverse that outcome. Village hardship and combined villager-official resistance notwithstanding, the Office evidently found Kiso too important to concede.[6]

2. *Leasing: Its Problems and Its Uses*

Even where the Property Office accommodated local interests, the outcomes commonly rankled with both sides. While locals resented the new costs, restrictions, and complications that they confronted, officials in the Property Office found that the compromise arrangements – land leasing in particular – undercut the purpose of Goryōrin by reducing its fiscal value to the Household. Government foresters characterized the problems in land leasing this way: users generally make do with natural forest growth instead of developing plantation stands, which is wasteful; renters understate estimates of stumpage and its value; shared-yield stands also are under-reported and the rate of division brings little benefit to the imperial fisc; and secret felling continues.[7]

As a whole, during the 1890s, the strategy of accommodation allowed appreciable areas of Goryōrin to be leased out, especially in areas such as Tōhoku where prefectural authorities continued to administer them until 1897. With those precedents established, however, in later years the Property Office faced severe difficulty dissolving such arrangements. And that situation undercut its efforts to implement the *hōseirin* policy of scientific forestry. Instead, it eventually ended up disposing of much of the leased acreage.

Leasing led to even more complete timber losses when it shifted land into agricultural use, and both leasing and the conversion of woodland to pasturage and tillage continued into the 1920s. Despite earlier sales and disposals, as of 1910, by one calculation, 116,000 *chōbu* of *goryōchi* (about six percent of the total) were still under lease, divided roughly in thirds between pasture, arable, and brushwood/bamboo growth. In following years, new leases were approved faster than unwanted parcels were disposed of, especially in Hokkaido, and leaseholds grew, peaking in 1924 at 134,865 *chōbu*. Nearly half of that acreage was in cultivation, mostly as dry fields.

There were, however, important regional differences between land leasing in Naichi and in Hokkaido. In Naichi, many leaseholds dated to the nineteenth century or earlier and were accepted reluctantly by Goryōrin officials when they first received their parcels of land. In Hokkaido, by contrast, the Property Office initially promoted leasing as a way to create a resident population sufficient to meet its forest labor needs and sustain a transportation system. And it continued that policy through the 1910s.

For example, in northwest Hokkaido major parcels of Goryōrin flanked the upper reaches of the Teshio River, which descends the western face of Mt. Sanru and then snakes its way northward past Nayoru (see Map 7). Starting in 1899, government authorities opened the Teshio Plain to development, and for several years thereafter Household officials used leases as a way to attract settlers. They set rental rates at roughly half that of nearby farm land and encouraged the settlers to use their winter months logging. Initially that work opened more leasehold land to farming, but subsequently it became a way to generate added wages for the farmers and market income for the Imperial Household. By 1917, the program had transformed 6,038 *chōbu* (roughly 17%) of the Sapporo regional headquarters' large Sanru operations area into agricultural leaseholds and turned the Teshio into an active artery of timber transport.[8]

Similar policies in other operations areas resulted in extensive and continual leasing of Goryōrin in Hokkaido until around 1920. By then, however, transport lines were better established and labor more available, which undercut the basic rationale for leasing. So in both Hokkaido and Naichi the Property Office sought alternative means of responding to local needs, hoping thereby to facilitate the elimination of leaseholds.

3. *Phasing Out Leaseholds*

One particularly notable policy that the Property Office adopted in 1920 appears to have been intended to reduce the demand for leases and other use arrangements while addressing a broader issue of social equity and village discontent.

To explain, for decades the government had been pressuring villages to establish schools, town offices, roads, and other municipal facilities and to defray such costs through local taxation. Mountain villages surrounded by *kokuyūrin* and Goryōrin complained that they were unfairly burdened because those woodland properties were tax-exempt and could help them cover local civic costs only insofar as they were leased or otherwise made available for villagers' use. In June 1920 the Property Office responded to this grievance by adopting a policy of giving villages payments-in-lieu-of-taxes for the Goryōrin acreage within a village's boundary. The level of payment was based on the estimated rental value of the property, as indicated by nearby rentals of comparable land, whether woodland, waste, dry field, or paddy field.[9]

This policy of payment-in-lieu may have lowered demand for new leases at what proved to be a very modest cost to the Imperial Household, but it didn't end existing leases. Other developments, however, did have that effect. In the early twenties, Japan's agricultural economy began slipping into recession, and as the decade advanced, the rural slump dragged on, sharply worsening during the Great Depression, 1929–33. With rural hardship widespread, the idea of selling leased parcels to their holders on easy terms as a way to assist the hard-pressed gained favor. And policy followed suit. The disposal of leaseholds accelerated in both Hokkaido and Naichi, as we note more fully below, with the Property Office selling parcels in a manner intended to ease local hardship while improving the quality of the remaining Goryōrin. By 1937, the policy had reduced

Table 3-1: Types and Acreage of *Goryōchi*, 1940 (in *chōbu*)

Category	*Hereditary*	*Regular*	*Total*
Palaces	451	578	1,029
Tombs	–	642	642
Woodland	208,511	1,062,394	1,270,905
Farmland	–	39,391	39,391
Residential	37	189	226
Miscellaneous	64	4,228	4,292
Total	209,063	1,107,422	1,316,485

(Source: Kuroda Hisata, *Tennōke no zaisan*, p. 112.)

leaseholds from the nearly 135,000 *chōbu* of 1924 to about 50,000. And sales continued in succeeding years. By 1940, as Table 3-1 indicates, *goryōchi* contained 96.5% woodland, 3.0% farmland, and 0.5% palaces, tombs, and similar.[10]

THE COURSE OF SHRINKAGE

A complex interplay of forces and factors thus shaped, complicated – and ultimately thwarted – the Property Office's attempt to form a large and coherent, Fuji-centered Goryōrin. Looking at the issue in terms of chronology, during the 1890s, as noted above, the Imperial Household lost a large amount of land in Hokkaido. It also released a smattering of parcels in Naichi, some in response to requests from other branches of government and a few in response to civil litigation. And, in their Yamanashi holdings, Household officials encountered difficulty almost from the day those were acquired in 1889. There, as we see in Chapter 4, their best efforts notwithstanding, they watched affairs worsen over the next two decades, finally abandoning the entire prefecture in 1911. And elsewhere they compromised their own interests on many parcels and sold many others, doing so from around 1900 on through the 1930s.

Legalities and Legislation, 1896–99

During 1898–99 the Household Ministry attempted to seize the initiative in woodland disputes by developing a policy that would enable it to control the decision-making process. It did so by adopting procedures very similar to those being devised for *kokuyūrin*.[11]

To explain, during the nineties both the Imperial Diet and Supreme Court had taken steps that aided villagers in disputes with government foresters. Shortly after the Diet assembled, it formed committees to study forest issues and hear the views of concerned constituents, notably landowners and lumbermen. It then passed pieces of legislation that gave villagers grounds for challenging government decisions on use rights, woodland rentals, and related matters involving *kokuyūrin*. Most importantly, a new Civil Law Code, approved by the Diet and issued in 1896, affirmed the legality of *iriai* rights. Using that law, villagers pursued litigation to reclaim such rights on *kokuyūrin*.

The Supreme Court, meanwhile, had ruled in the plaintiff's favor in a number of minor woodland cases. And, in May 1898, in its most important decision regarding forest, it ruled, on the basis of the 1896 law, that decades of government assertions had been in error and that villagers did indeed have a legal right to *iriai* yields, even in *kanrin*.

These legal developments opened the floodgates to litigation, compelling the government to devise procedures for settling disputes on *kokuyūrin* in an orderly manner. The Forest Agency proposed a procedure for handling petitions and litigation, but during long-drawn-out consideration in 1898–99 a testy Diet committee subjected it to sharp criticism and forced the Agency to accept modifications. Finally, in April 1899, the Diet approved the bill, which spelled out how villagers were to petition for recovery of claimed rights and how they could appeal failed petitions through the judicial system.

This body of legislation and adjudication had no direct legal force in Goryōrin, which, as noted earlier, was perceived as neither *kokuyūrin* nor *min'yūrin* but as a third category of woodland.[12] Nevertheless, the problems it was addressing in *kokuyūrin* also afflicted Imperial Forest, and now that villagers could seek redress through the courts in the case of government woodland, the Household Ministry felt compelled to devise orderly procedures of dispute resolution for its own estate.

Household Ministry Maneuvers, 1899–1901

The Diet legislation of 1899 presented the Ministry with a fundamental conundrum: could the Emperor, the very Fountainhead of Justice, who had bequeathed the Meiji Constitution to His People, have His Property Rights in Goryōrin challenged in the very Courts that His Constitution had made possible? Ministry officials decided, somewhat tentatively, that he could not (see Appendix C). So they had to find some way to address villagers' complaints before they went to court, where the Emperor's immunity might be challenged and His Benevolence thrown into question.

It was a reasoned decision, but it carried some unintended costs. Most directly, it closed off the courts as a peaceful and binding legal venue for dispute resolution. As an ironic consequence, in 1915, when the Supreme Court, in a case involving a village in Nagano Prefecture, reversed its 1898 position and ruled that villagers' claims to *iriai* use rights on *kokuyūrin* had no legal force, the Household Ministry was unable to apply that ruling to Goryōrin to justify consolidation of its control.

Having decided that it must sidestep the new judicial system, the Ministry tried to resolve disputes over woodland use by adopting its own version of the petition procedure recently established for *kokuyūrin*. In May 1900, it announced that it would return *goryō* woodland and standing timber without charge to anyone who could document his claim to it in a petition submitted to an evaluation committee within the Property Office. Such applications must be submitted within a year, however – by June 30, 1901, to be precise. And in April 1901, by which time it was clear that many petitions laid claim to parcels of *seden goryō*, the Ministry announced that although hereditary parcels could not formally be returned, it would allow contractual leasing and harvesting activity on them. These moves, it was hoped, would obviate recourse to the courts.

Public Responses, 1901–18

During the year of grace the public responded to the Household Ministry's offer with stunning vigor. Villagers submitted petitions that challenged its claim to some 490,000 *chōbu* of Goryōrin on Honshu, about a third of the Naichi total. The petitions numbered 1,800, and of them the evaluation committee concluded that 1,076 merited site inspection.

In April 1902, while these cases were in train, the Household Minister secretly instructed the Property Office's head to be generous in his settlements so as to display imperial benevolence toward the petitioners. However, his advice, even if sincerely given, went unheeded, probably because the legal definition of what constituted evidence of a legitimate claim was so strict. Also, the committee may well have feared to open the floodgates, given the magnitude of claims submitted. And in any case, it was an instrument of the Office, whose mission was to preserve, not fritter away, Goryōrin. Whatever the determinative factors, of those 1,076 petitioners, only twenty-eight received parcels of land, totalling 967 *chōbu*, while eight others received standing timber or bamboo. The imperial benevolence had proven rather thin.

In similar cases involving *kokuyūrin*, where the rejection rate was also high, the legal system allowed failed petitioners to appeal to administrative courts. But on Goryōrin, petitioners were strongly discouraged from legal recourse. Some thirty suits were in fact filed in civil court over Goryōrin issues, most involving disputes over particular boundaries or ownership claims, and in them the Property Office or Household Minister, not the emperor, was named defendant. In nearly all instances, however, the defendant prevailed, the last cases ending sine die in 1918.

For all practical purposes, then, in Goryōrin matters the courts were closed to village plaintiffs. This being so, many concluded that their only real choices were to seize what they could, use the land illicitly, or damage it enough so that the Property Office would decide it was not worth keeping and dispose of it "by choice." Villagers did in fact employ all three strategies, and all three proved useful.

Disposal "by Choice," 1898–1937

Regarding disposal "by choice," almost from the outset builders of Goryōrin (as well as *kokuyūrin*) had realized that they were getting a poorly-surveyed miscellany of lands and that many pieces were hardly worth keeping. It was not until January 1898, however, that the Property Office felt ready to take the initiative in disposing of unwanted parcels – by which time the Imperial Household had expanded its trained forestry staff, completed substantial resurveying, set up more regional offices to manage woodland, and begun working on its first general management plan for Goryōrin.

That month (even as the Diet was discussing forest law and administration for *kokuyūrin,* querying en passant its relationship to Goryōrin, and even as the Forest Agency was developing its own plans for disposing of unwanted properties and managing the rest) the Property Office enumerated criteria for its regional foresters to use in sorting out their land. Properties were to be retained if they appeared to have value as timber-production forest or as protection forest for safeguarding waterways, roads, fisheries, and such against wind and flood. The parcels had to be large enough to be useful. And they had to be accessible, not entangled with patches of *min'yūchi,* and not hobbled by existing legal commitments.[13] Properties that failed to meet these criteria were candidates for sale.

The Property Office's primary objective, which was very much in accord with contemporaneous planning by the Forest Agency, was thus to dispose of small, scattered, and compromised parcels. The monies it received would then be used to defray costs of administering and revitalizing the woodland it wished to retain and develop as hereditary *goryō.* And indeed, to foreshadow the outcome, the effort did pay off. Between 1889 and 1937, by one calculation, sales of *goryōchi* produced income of nearly 43,000,000 yen, mostly during the 1910s and twenties. And prior to 1914 that income had covered more than a third of the Property Office's total annual expense on forest operations (see Table E-13).[14]

That profitable outcome notwithstanding, the process of selling parcels, which occurred in three stages, proved cumbrous and time consuming.[15] The first sale, announced in 1898, soon got snarled up in a flurry of complications and policy modifications. But after a pause it gradually made progress, and by 1911 it had (in rounded figures) disposed of some 12,000 *chōbu* scattered in 8,500 parcels, mainly west of the Kantō. Most of the land was sold to those villagers who were already using it in one way or another. A separate 33,000 *chōbu* in 9,000 small parcels were sold to the fanes that had held them prior to their seizure by the government back in 1871.[16]

By 1908, the Property Office had identified another assemblage of parcels that it deemed undesirable, and that year it announced a second sale, even as the first was nearing completion. In a heightened effort to assure that villagers, rather than outside speculators, would acquire the land, this sale was explicitly restricted to local people with existing ties to the parcels. And since many were people of modest means, buyers were allowed to pay in installments for up to ten years. By 1926, when that program was completed, it had disposed of another 11,000 *chōbu* in 3,900 parcels.

The third and final sale was substantially larger. It began in 1919 as a planned disposal of some 51,000 *chōbu* of unwanted agricultural land that the Property Office had been leasing out. But then in 1925, in apparent response to the widespread rural hardship of the day, which we noted above, it was expanded to include woodland as well.[17] The Office then designated for sale those sites carrying *iriai* rights, pasturage, and *buwake-bayashi* agreements, along with areas that otherwise showed no prospect of producing good timber. And again it tried to restrict sales to locals, including villages and towns. By 1936, this third sale had disposed of about 198,000 *chōbu* of farm and forest land, and the Property Office had identified another 60,000, mostly leaseholds, that were available for sale if buyers could be found.

In sum, then, not counting that final 60,000, over the course of about thirty-five years the Property Office had initiated the disposal of some 254,000 *chōbu* of *goryōchi*, mostly by sale to local buyers. However, because the parcels it sold were generally land that villagers believed they had a right to use in any case, many resented having to pay to recover what had once been theirs more or less gratis. But they did pay, in many cases doing so only because they had previously failed to pry the land out of Property Office hands by other means.

* * * * *

From first to last, then, during its brief lifetime Goryōrin never achieved complete stability of size, location, or character. Ever shaped by the social context within which it was created and developed, it was continually subjected to outside pressures and gradually shrank, whether due directly to external duress or as a result of administrative choice. Most of the elements in this contentious process of sorting out Goryōrin are nicely illustrated by the case of Yamanashi. That case ended up costing the Property Office an additional 298,237 *chōbu* of woodland. And more than any other development, the Yamanashi fiasco destroyed the vision of a Fuji-centered *goryōchi*.

4

The Case of Yamanashi Goryōrin

To summarize the story of Yamanashi Goryōrin, the problem there was shaped fundamentally by the prefecture's geography and the local socioeconomic circumstances fostered by that geography. It was also shaped by the *kokuyūrin* policies pursued there during the 1870s and 1880s, which underlay Property Office decisions of the 1890s, and by the response of villagers to those government moves, which led to continual modification in both *kokuyū* and *goryō* policies.

The net result of four decades of this thrust and parry was serious damage to the prefecture's woodlands. Despite some successful efforts at reforestation and other forms of remediation, erosion, landslides, and floods worsened, culminating in hugely disastrous inundations in 1907 and 1910. Those calamities finally led a dispirited Household Ministry to abandon its whole effort. In 1911, it turned over to Yamanashi prefectural authorities nearly 300,000 *chōbu* of woodland, of which 79,300 had been supposedly sacrosanct *seden goryō*. And with that loss the vision of a Fuji-centered imperial estate was irreparably compromised.

GEOGRAPHY

Looking briefly at the geography, Yamanashi Prefecture occupies some 450,000 *chōbu* (nearly 4,500 km^2) in the southeastern quadrant of Japan's sharply incised central cordillera. Except for the northerly slope of Mt. Fuji, which lies along its southern flank, and the lovely but small (roughly 20,000 *chōbu*) triangular Kōfu Basin, the prefecture consists of a maze of acutely upthrust ridge lines separated by the rushing waters of two highly articulated, dendritic river systems (see Maps 3, 4, 6). Most of Yamanashi's eastern quarter drains eastward via tributaries of the Sagami River, which debouches onto the Kantō Plain, thence flowing into Sagami Bay near Hiratsuka. The other three quarters of the prefecture drain southward via tributaries of the Fuji River, which empties into the upper end of Suruga Bay.

Being an assemblage of narrow, twisting – and beautiful – mountain valleys, little of Yamanashi's land surface (about eleven per cent) is arable.[1] For centuries those sinuous valleys sustained a sparse and comparatively poor population of villagers who eked out a living from the harvest of dry fields and miscellaneous forest products and by working away from home (*dekasegi*). That very sparseness of settlement in turn helped assure that Yamanashi remained well forested and hence attractive to the builders of Goryōrin.

Yamanashi woodland was capable of growing fine stands of merchantable *sugi* and *hinoki* thanks to its topography and climate. And being situated only about 100 km from central Tokyo and less than 200 km from Nagoya, it was relatively close to major wood-consumption centers. However, its river systems were not convenient for floating or rafting because their streams were so swift, shallow, ledgy, and boulder-strewn. So timber transport was exceedingly difficult, and relatively little entrepreneurial lumbering had ever developed there. Even government-sponsored logging was spotty, and mostly it entailed the extraction of smaller-sized products for market or local use. So, as of 1870, the prefecture's hillsides mostly continued to support mixed stands of old-growth forest.[2] For people versed in German forestry, they must have seemed rich with silvicultural potential.

EARLY MEIJI DEVELOPMENTS

During the 1870s, as noted earlier, Meiji leaders undertook the land surveying and *kan/min* division that distinguished government property from that of others. As the project's difficulties became clearer, leaders in Tokyo tightened their control, requiring prefectural officials to obtain Ministerial guidance in ambiguous cases. By the mid-seventies that requirement was producing a vast correspondence in which prefectural governors were attempting to explain to Tokyo the particulars of specific cases involving as little as a few acres here or there, a hillside or two, or one or two villages. In some cases prefectural authorities were simply seeking guidance; in others they wished reinforcement for unpopular actions; in yet others they were defending local interests against central authority.

In Yamanashi, most of the lowland areas, mainly in the Kōfu Basin, were quickly identified as taxable *min'yūchi*. And some of the most mountainous areas easily qualified as *kan'yūchi*. But surveyors found that villagers claimed *iriai* use rights on more than seventy five percent of the prefecture's upland, and because of that complication they were unable to complete their official examination of its status until 1881. Then, however, they ruled that of this *iriaichi*, only 3,385 of 352,888 recorded *chōbu* qualified as *min'yūchi*; the remainder was government land.[3]

This *min'yūchi* figure was so low because so little of the woodland had a documented record of actually being managed by villagers in a sustained fashion. In fact, villagers had been utilizing those hillsides for generations, mostly through common-use arrangements and by the customary method of exploitation forestry: take what you need and let the forest restock itself. They viewed the land as theirs to use for little or no fee and in the manner to which they were accustomed. And they continued doing so despite the new designation of those hillsides as *kan'yūchi*, which from their perspective amounted to government theft pure and simple.

Indeed, in some areas local dependence on woodland had increased substantially in recent decades. A booming market in silk exports developed during the 1860s, and from then on villagers in some parts of Yamanashi used more and more hillside for growing mulberry to feed the silkworms.[4] The hillsides also produced fuel for heating brood trays and the water used in extracting silk thread, as well as fodder to feed the swelling number of horses that carried the silk to market. This newly expanded market opportunity enabled more people to live in the region and improved their lives. But in the process, it increased their dependence on the areas newly designated as *kan'yūchi*. When officials attempted to stop such use, villagers resorted to illegal harvesting, theft, and arson, thereby precipitating more erosion and downstream damage.

During the Matsukata deflation of the early eighties, villagers in central Japan became particularly resistant to trends of the day, and the disorder spurred prefectural officials to lobby Tokyo leaders for concessions. The governor of Yamanashi, alarmed by growing public disorder and environmental damage, wrote petitions on behalf of his constituency. In a proposal that amounted to de facto restoration of *iriai* arrangements, he recommended to Tokyo, in May 1882, that villagers who plant tree seedlings on *kokuyūrin* be given fifty-year leases (*nenki*) on their handiwork, while those who simply wish to harvest self-seeded trees or other growth receive twenty-year leases.[5] He claimed that surveyors had identified 182,800 *chōbu*, fully half the prefecture's former *iriaichi*, where such arrangements would be appropriate.

The Ministry of Agriculture and Commerce (Nōshōmushō) replied later in the month, approving the twenty-year leases on harvest of volunteer growth. But it applied the same twenty-year limit to plantation work, thereby undermining the incentive for afforestation. That change led to a flurry of exchanges, which culminated in January 1883 with the prefecture's issuance of an elaborately detailed thirteen-clause set of regulations on the use of *kan'yūchi*.[6]

Those prefectural regulations restored the difference between exploitation and regenerative practices, but they did so with sharply reduced time limits: twenty-year leases for tree planters and five-year leases for those who wished simply to cut trees, brush, and grass. Moreover, fees would be charged. And other stipulations sought to protect woodland from abuse by spelling out acceptable harvest procedures, fire-control requirements, and other restrictions on forest work. The arrangements were cumbersome, and villagers resented the attempts to infringe on what they regarded as time-honored use rights. Some agreed to follow the new procedures, but others did not. So damaging overuse continued, which led during the next few years to not very effective tightening of the rules on woodland use.

Long before the Imperial Household Ministry acquired Yamanashi woodland, that is to say, it was already hotly contested ground.

GORYŌRIN: EARLY PROBLEMS

As government surveyors pursued their work during the eighties, their estimates of prefectural forest acreage slowly became more accurate. By 1889, the government claimed to have in Yamanashi some 415,000 *chōbu* of woodland, of which 308,704 were *kanrin* under Forest Agency authority, while an additional 106,255 were *kan'yū rin'ya* under prefectural control. That August, it transferred the land and stumpage in both categories to the Household Ministry, thereby giving it most of the prefecture's woodland.[7] However, prefectural officials still administered the land, so they continued to face the familiar difficulties of woodland management, notably local disregard for leasehold arrangements. But they now were supposed to deal with them in line with regulations coming from the Property Office's regional headquarters in Shizuoka, which had been newly established that month to oversee the Household's Fuji-area woodland.[8]

The Matter of Leases

By then many of the villagers' initial five-year leases of *kan'yūchi* were nearing expiration. Fearful of losing their harvest rights because of the land's conversion to Goryōrin, they intensified their felling activity. In March of 1890, a forester from the Shizuoka headquarters sent the Property Office a report urging that the leases be renewed so as to alleviate the fear and stem the damage. He described the procedure and problems of logging in Yamanashi this way:

> As per the regulations for selling [stumpage], parcels where stands may be sold are designated [by foresters] annually, except in protection forests. Even though felling is authorized only within each parcel's boundaries, however, chopping is actually pursued inside and out without distinction. Or else people expand the acreage of stumpage they buy every year and, having recklessly bought so much, flood the market. Finally, fellers willingly leave defective trees, branches, and so forth in the forest, so that eight or nine out of ten pieces brought out are high-grade trunk-wood. Although the people are charged with protecting forest cover, it is the responsible persons themselves who do the reckless cutting, resulting in this pathetic situation.[9]

He then went on to urge that leasing arrangements be extended to eliminate the pressure to cut so hurriedly. But he also recommended the inclusion of more protective regulations to prevent further abuse.

The Household Ministry accepted the recommendation and shortly acted upon it.[10] The Property Office drafted a new set of regulations, basically following the older Forest Agency version. Except that it designated the Shizuoka regional headquarters as the office in charge, allowed use rights in perpetuity rather than for a fixed period, and expanded by about a third, to 240,000 *chōbu*, the acreage on which communal use would be allowed. The Office, it appeared, was going to be far more responsive to local needs than the Forest Agency had been. As usual, however, the devil was in the details.

Shizuoka was nominally the office in charge, but Yamanashi prefectural officials still administered the land. So, in September 1890, the prefectural office issued an eighteen-clause version of the new Property Office regulations. The Yamanashi rules specified that those villagers who applied for contracts could secure use rights in perpetuity. New felling fees were to be established after a five-year grace period, however, and adjusted annually thereafter. Site restrictions were to be enforced; users were obligated to restock areas; and miscreants were to be fined as per schedule. Villages and common-use associations (*iriai dantai*) were to be held communally responsible for submitting – via village heads – formal requests to harvest. When making those requests they were to use the approved forms, prepare maps of the sites, specify their rotation-cutting schedules, and secure Shizuoka's approval before starting to harvest. They were to assure that rotation-cutting rules were properly followed, and they had to complete each cut within a year of its authorization.[11]

From a forester's perspective, especially one inspired by the new silvicultural thought coming from Europe, these were thoroughly sensible rules. But to most villagers they were thoroughly irritating. Most villagers, after all, were illiterate. But they were not fools, and few were eager to commit themselves to elaborate agreements they could not even decipher. Even where elemental trust existed, the stipulated procedures were cumbrous, time-consuming, and costly – all that wretched bother simply to work hillsides that "we" had used for generations through communal agreements, without outside meddling, and at little or no charge.

Less obviously, the "village heads" through whom applications must pass were not the heads of "real" villages, meaning the long-established agricultural hamlets. Rather, by 1890 those hamlets had been redesignated *buraku* and subordinated to new and larger administrative units, the *mura* or "administrative village."[12] The head of such a *mura* was very likely a stranger with whom most locals had no contact except through their *buraku* representative in the recently established village assembly. And that assembly was composed of the *buraku* elite, who might give short shrift to the interests of poorer neighbors.

In consequence, the Household Ministry's generous concessions on leasing elicited little enthusiasm, and few villagers even bothered to apply for contracts. Instead the reckless and illegal felling continued, prompting officials in 1895 to simplify policy. Revised regulations abandoned the rotation-harvest scheme and tried closer oversight, use of permits, and the adoption of shared-yield (*buwakebayashi*) arrangements in hopes those would improve compliance.[13]

Villagers' Use of Goryōrin

Household foresters had two basic silvicultural objectives: to establish protection forest where it was needed and productive timberland elsewhere. The basic objectives of villagers were rather more complex and fluid. They wanted to avoid damage to hillsides, certainly, but they also had to make a living by one means or another. In essence, they were trying to reconcile concurrent desires to market silk, work as packhorse operators, and take advantage of a burgeoning timber market. And the three did not mesh smoothly.

From the outset, of course, some villagers did try to follow the new rules for leasing woodland, with hamlets and *iriai* organizations drawing up and repeatedly revising rules to guide their members in exploiting the acreage they were allowed to use.[14] During the 1890s, for example, the hamlet of Nagaike, on the north shore of Lake Yamanaka at the edge of Mt. Fuji's lava skirt, regained access to nearby areas of brush and woodland that it had used previously. Residents planted mulberry bushes on the land, expanding their acreage, which was listed as 54 *chōbu* in 1888, to 565 by 1904. And they continued using the land for diverse other purposes as well.[15]

Of these other purposes, the most vexatious proved to be the pasturing of horses. Villagers used horses as pack animals for regional transport, which constituted a major and growing form of employment. Already by 1886 Nagaike had sixty-three horses stabled in thirty of the hamlet's thirty-four households, and by the mid-nineties free grazing by the growing herd was damaging so much land that hamlet regulation of the practice gradually tightened. *Buraku* rules of 1896 prohibited all free grazing, and those of two years later were even more restrictive. They specified areas where tethered pasturing would still be allowed but others where even that was forbidden. And they detailed the fines and other punishments for infringement.[16]

One reason grazing became such a problem was that as more land was logged off or converted to mulberry culture, the extent of erosion and flooding increased, and grazing horses only exacerbated the problem. Another objection to grazing was that pastured horses trampled both mulberry shoots and tree seedlings. And when other forage grew scarce, they nipped them as well, reducing mulberry yield and usually ruining a seedling's potential as a timber tree. Tethered grazing could protect mulberry plantings, but not the tree seedlings that sprouted randomly on exposed hillsides. Because the demand for construction timber was boosting prices as the 1890s advanced, villagers had ever more incentive to log and, therefore, to grow new crops of trees. So they increasingly objected to grazing on any areas that contained tree seedlings, whether planted or self-seeded.

Moreover, the means of transport out of Yamanashi was changing, making the lumber market more and more accessible. Specifically, in 1896, workmen began extending the Chūō Railroad westward from Hachiōji, and the line shortly entered Yamanashi via the Sagami River valley. It came within thirty kilometers of Nagaike at Ōtsuki, and continued westward from there. By 1904, after arduous digging and blasting to form the three-mile Sasago Tunnel, the line reached Kōfu and opened for business to that point.[17] Its operation reduced sharply the cost of shipping lumber and silk from Yamanashi to Tokyo, giving villagers in the Nagaike vicinity further incentive to fell accessible trees and ship out the wood. They could then plant the clearings to mulberry, or to a new timber crop if the terrain was too inconvenient for such use. Those practices enabled them to produce more and more goods that they could move by packhorse and cart or sled down to the railroad siding at Ōtsuki for easy and rapid shipment into the city.

The growing interest in lumbering, and hence in reforestation was suggested by changes in Nagaike's *iriaichi* regulations. Most notably, its 1908 version included new clauses detailing procedures for reforesting parcels of *iriai* land. Precisely why that change was made is unknown. But it was a fair reflection of several trends: the growing appeal of timber marketing; heightened pressure from high authorities to pursue such measures; and the gradual spread of fashionable silvicultural thought. But perhaps it was a response, most basically, to the widely-felt need to cope with the natural devastation that had become more pronounced during the 1890s–1900s and that was by then ravaging much of the prefecture.

GORYŌRIN: LATER DEVELOPMENTS

By the late nineties, the Property Office had made good progress in establishing control over its Yamanashi woodland. In the process, however, it had discovered that its acreage was substantially less than original estimates had indicated and that its condition was worsening rather than improving. By then, others too, villagers most notably but also prefectural authorities and lumbermen, were aware that troubles were escalating and that something had to be done about it.

Government Measures

During the nineties the Property Office's Shizuoka headquarters gradually acquired the staff necessary to oversee Goryōrin in neighboring Yamanashi. By late in the decade, therefore, it was taking charge of the more detailed surveying that was identifying boundaries, acreages, and stand composition ever more precisely. As survey reports reached Tokyo, the Property Office staff became more and more aware of how seriously damaged much of the woodland was. In response, they drafted plans to rehabilitate deteriorated sites, proposing to establish protection forest on 34,377 *chōbu* during the next two decades. In 1898, the Household Ministry instructed villages to begin local projects in that program.

Meanwhile surveying continued, and a prefectural report of 1903 tabulated the results as in Table 4-1. What had been called 308,704 *chōbu* of *kanrin* in 1889, and been viewed as incipient *seden goryō* for the Household's Fuji-centered estate, was restated here as 84,900. And eight

Table 4-1: Yamanashi Goryōrin in 1903 (in *chōbu*)

Forest Category	Register Acreage	Percent of Total	Verified Acreage	Notations
Previously *Kanrin*	84,900	20.8%	52,600	"pure" *goryōrin*
Surrendered by fanes	23,100	5.7%	14,400	may be returned to fanes
Iriai goryō	299,600	73.5%	185,800	may be returned or held with *iriai* use
Total	407,000	100%	252,800	

(Source: Tsutsui Michio, *Nihon rinseishi kenkyū josetsu*, p. 102.)

years later, in 1911, as we note below, it was listed as 79,300 *chōbu.* Reflecting the influence that former forest holders had regained during the nineties, some 23,000 *chōbu* taken from shrines and temples back in 1871 were re-classified as lands that might be restored to them. Most importantly, where surveys of 1882 had found 182,800 *chōbu* of land with *iriai*-type uses (a figure restated in 1890 as 240,000), the acreage was recorded in 1903 as 299,600. And in line with recent moves to sort out woodland and eliminate undesirable parcels, this expanded *iriai goryō* was listed as subject to potential disposal. In all categories, however, the actual acreage may have been less, as surveyors had still been able to assess and catalogue only about sixty per cent of their presumed holdings.

As the mere existence of this prefectural analysis suggests, the Household Ministry was not the only body troubled by worsening conditions. Even as Yamanashi authorities were ceding control of woodland to the Property Office's Shizuoka regional headquarters, forest deterioration was creating ever worse social and fiscal problems for the prefecture's people and officialdom.

During sessions of the prefectural legislature in 1900 and 1901 members discussed the issue and prepared a scathing report. In short paraphrase, they warned that ever since the restoration, Yamanashi forests had been abused. So fuelwood and lumber have become insufficient; flooding and erosion are ruining terrain and people's productivity; the costs of riparian repair are escalating, and year after year the regional economy deteriorates. As for the future, if the forests are not repaired, flooding will worsen and the prefecture's people will be devastated.[18]

At the end of 1901, legislators prepared a rough estimate of the cost of needed riparian work. They noted the difficulty of managing woodland effectively and asserted that all of the prefecture's resources must be directed to the problem. They detailed the ways in which the forest problem was damaging Yamanashi: in lost income; in losses of irrigation water and fertilizer materials; in deaths, injuries, and bereavement; in the costs of enforcing forest law and prosecuting violators; and, finally, in the cost of repairing and replacing river banks and dikes, ruined roads and bridges, and wrecked fields and irrigation systems. They included cost estimates for these several items and proposed ways to improve woodland, mainly by better silvicultural practices. But how to implement the ideas they did not say, and in a despairing conclusion they declared that they did not know how to cope with the situation.[19]

The tone of despair notwithstanding, prefectural officials set to work drafting regulations to improve policy in what they called "protection forests" (*hōanrin*). And, in August 1902, the governor issued them as a twelve-clause set of prefectural rules.[20]

In a manner comparable to British forest policy of the day, these rules completely disregarded the formal ownership categories of private, government, and imperial forest. Instead, they treated the prefecture's woodland in terms of two basic types: those areas where logging was simply prohibited (called *hōanrin*) and those where it was not (called *futsū hōanrin* or "regular protection forest"). They then spelled out, for each type, management procedures and restrictions on use. Where felling was not prohibited, the rules specified how to proceed when proposing to log, detailing the particulars of felling, replanting, and reporting procedure. And they indicated the site conditions that would require prefectural permission before any work at all could proceed.

Where felling was prohibited, the rules spelled out the conditions under which other activities could proceed. These activities were the customary rental and *iriai* uses that villagers made of both government forest and Goryōrin: the thinning of stands, removal of dead or damaged wood, emergency cleanup work, removal of soil, rock, or grassy undergrowth, digging up of roots, pasturing of animals, and use of fire for clearing useless undergrowth. Finally, the rules stipulated the levels of punishment (fines) for those convicted of particular violations.

The following year, Yamanashi authorities drafted an ambitious proposal for a ten-year program of forest restoration through collaboration with villagers.[21] They called for education and funding assistance that would enable local folks to reforest deteriorated sites and regulate the use of other areas. Their proposal, with its attention to public participation and support, reflected the growing influence of more well-to-do landowners, whose voice was being heard through both the legislature and the administrative villages that stood between hamlet and higher authority.

In this prefectural response to the problems of disintegrating woodland, one sees what might be called the triumph of biology over politics. Whereas prefectural notices of the early and middle 1890s had explicitly addressed issues of *goryōchi* as an independent category of land, by 1898 they were speaking explicitly about issues in "both *goryōchi* and *kokuyū rin'ya.*" Since the rules of *kan/min* division had tried to place under government control all lands susceptible to erosion, landslide, or other communal danger, the shift in terminology – culminating in the implicitly universal "protection forest" regulations of 1902 and the ten-year collaborative program of 1903 – suggests that policy makers recognized the irrelevance of political categories to the environmental crisis they were facing. In this Yamanashi approach to woodland management, one sees a foreshadowing of both the later calls (mentioned earlier in Chapter 3) for recombining *goryō* and *kokuyū* forest and the vigorous moves of the 1920s–1930s (which lie beyond the scope of this book) to extend government guidance into the operation of private woodland.

Villagers' Measures

At yet more local levels, as well, forest deterioration was provoking action of diverse sorts.

Most constructively, perhaps, a number of hamlet *iriai* groups responded to the Household Ministry's 1898 call for creation of protection forests by offering to handle the planting as shared-yield (*buwakebayashi*) projects, in line with the Property Office's revised regulations of 1895. Doubtless the villagers were aware of the booming lumber market and judged the long-term profit outlook worth the immediate cost of nurturing the areas contracted, especially because it would also give them tax-free use of underbrush and thinnings even as it stabilized vulnerable terrain. In any case, the Shizuoka headquarters (and later a short-lived Kōfu regional headquarters that the Property Office established in 1906 in its faltering effort to salvage its Yamanashi holdings) accepted the offers, and shared-yield planting intensified. Whereas there were only 155 *chōbu* of such plantings in the prefecture as of 1888, by 1910 an additional 7,025 *chōbu* had been contracted.

Other hamlets and *iriai* groups responded differently. Most aggressively, some petitioned to acquire complete ownership of parcels of Goryōrin. Thus in April 1898 (even as the plans and proposals, noted in Chapter 3, for the sale or surrender of *kokuyūrin* and Goryōrin were circulating among government offices in Tokyo) the hamlet of Nakano, near Nagaike, submitted a petition requesting the Property Office to return parcels of woodland that it had been trying to recover since 1883.

Two years later, in August 1900 – in response to the Household Ministry's May offer to return land gratis to any who could prove possession rights – the eleven-hamlet *iriai* association that included Nagaike and Nakano followed the latter's lead and petitioned to regain all of its erstwhile woodland. On June 17, 1901, two weeks before expiration of the Ministry's offer, the association submitted another, much more detailed, petition in which members carefully identified the sites, reported their stumpage, justified the request on grounds of long usage, and promised to pay taxes on the land once it was returned.[22]

Like nearly all petitions on that occasion, theirs came to nought. So on November 9, the association turned to the next best strategy, that of trying to make Goryōrin serve its needs. It submitted for Property Office approval a draft proposal containing fifteen clauses to guide its members in their use of a specified area of woodland. The clauses (see Fig. D-2 in Appendix D) reveal the similarity of these rules to those common during the Tokugawa period. But they also show the great effort villagers were making to sound as administratively punctilious as the Property Office. The clauses carefully spelled out harvest procedures, fees, restrictions, and fines and other punishments for violations, as well as measures to protect the forest from theft, fire, or other damage.

Although unclear, it seems likely that authorities provisionally accepted the fifteen-clause regulations. At least, in 1903 the eleven-hamlet association added two more stipulations. One specified that timber and fuel wood could be sold only within the village. The other noted that because the army had recently acquired the right to train and conduct maneuvers in the vicinity, villagers were forbidden to enter that area during such training.[23]

Finally, in early 1908, the Property Office issued a remarkably long and complicated set of regulations that constituted its final reply to the eleven-hamlet proposal. It spelled out in exhaustive detail how oversight was to be organized and implemented, how overseers were to be selected, and how all administrative procedures were to be handled.[24] Although it is difficult to imagine villagers adhering to all the particulars, the regulations did reveal the Office's tendency to manage with ever-increasing strictness. They also revealed, however, that here at the very heart of its Fuji-centered dominion, conditions were so desperate that it was prepared in the end to accommodate, however grudgingly, villagers' continued use of Goryōrin if they would help restore it to health.

CRISIS AND ABANDONMENT

All this time, during the 1890s and 1900s, while observers were debating what to do and villagers and officials alike were launching their initiatives and responses, the situation kept worsening. Widespread overcutting continued as villagers scrambled to meet the day-to-day needs of their farming, silk production, and other operations, exploit lumber market opportunities, and adapt to changing times.[25] Under the onslaught, woodland continued to deteriorate, and by 1907 a recorded 6,300 landslides had ruined some 4,800 *chōbu* of land in the prefecture.

As floods had worsened and damage mounted, more and more villagers had petitioned that land be returned to them. Finally, in June 1906, the Imperial Household developed explicitly for Yamanashi a plan to offer for sale, in line with existing administrative principles, all poorly-wooded parcels of less than 100 *chōbu*.[26] The regulations for that sale were as follows:

Regulations on Special Sale of *Iriai Goryōchi*

ITEM: Areas of Yamanashi Prefecture *iriai goryōchi* that have no trees: if no hope of trees and less than 100 *chōbu*, sell for fee to *iriai* associations.

ITEM: On such areas, determine sale price by simple survey of acreage and vegetation.

ITEM: After sales are complete, divide parcels in remaining Goryōchi to end *iriai* rights, or pay a fee to do so, or do so by the rule of future cost estimates. And where such rights continue, revise Yamanashi *iriai* rules.

In November, Yamanashi Prefecture publicly announced the sale, requiring that applications be submitted by the following June.

Household officials hoped, it appears, that, given the convoluted boundaries of Yamanashi *goryōchi* and the great proportion carrying *iriai* rights, a one-time sale would transfer enough land to Yamanashi villagers so they would be willing to surrender all *iriai* rights in most of the remaining Goryōrin, either accepting compensatory cash payments or dividing parcels between government and *iriai* users. With cash in and villagers out, it would presumably become possible to repair the surviving Goryōrin and thus preserve, at least in diminished form, their Fuji-centered forest realm.

To what extent the Special Sale was a response to reports from the Property Office's own foresters and to what extent a response to pressure from well-to-do landowners in Yamanashi is unclear. But within weeks of the sale plan's public announcement in late 1906, those rural leaders, who had thitherto been able to promote their interests through the prefectural legislature and the new administrative village system, created a new lobbying voice, the Yamanashi Sanrinkai, a prefectural Forestry Association of the sort that was appearing widely about the country.[27] Established in December and consisting mainly of entrepreneurial lumbermen and professional foresters, the Association's agenda called for forest improvement, the propagation of "scientific" forestry thought, and the development of forest industries, with lumbering as the core business.

Fearful, it appears, that wealthy outsiders might seize the opportunity to buy up large tracts of Yamanashi woodland, the Association urged that areas of *iriai* woodland be returned to their former users, not sold to just anyone. And it encouraged all hamlets to petition the Property Office for such return. To set an example, it petitioned for return of 200 *chōbu* that it proposed to rehabilitate.

In following months the Association remained active, expediting petitions and lobbying forest issues at both village and prefectural levels. Its effort was stunningly successful. But in a sense it also backfired. Petitions poured in to the Ministry, some 2,492 of them offering to buy 160,000 *chōbu*, roughly half the prefecture's non-hereditary *goryō*. Members of the Property Office staff, who surely shuddered at the prospect of losing that much land close to Fuji, were overwhelmed by the task of evaluating all the requests. As of 1911, even after nearly four years had elapsed, they had still assessed and approved the sale of only 894 *chōbu*.[28]

By then, however, natural disasters were completing a project that humankind had begun. By 1907, an estimated 90,000 *chōbu* of woodland – nearly a quarter of the prefecture's total – were severely denuded.

And during the typhoon season that August, rains and flooding that pounded the Kantō-Chūbu region did unprecedented damage in Yamanashi, ruining another 5,000 *chōbu* of protection forest. They also drowned 232 people, destroyed some 12,000 houses, and ravaged 775 *chōbu* of the prefecture's precious arable land.[29] By one calculation the typhoon's direct damage to Yamanashi cost 13,000,000 yen.[30]

Three years later, in a monster typhoon that wreaked havoc in seventeen prefectures between the Nobi region and southern Tōhoku – reportedly Japan's worst flooding catastrophe in a century – Yamanashi was inundated again. And again the havoc was massive. The 1907 disaster had unleashed newspaper attacks on government forest policy as injurious to villages, and this second assault in 1910 evoked even harsher denunciations of failures in government forest and flood-control policy.[31] The combined weight of that physical disaster, the public bludgeoning, and the seeming insolubility of Yamanashi's forest problems, when added to the huge burden of processing all those land-purchase applications, evidently broke the Property Office's will to persist.

That winter, the Household Ministry explored alternatives, and in March 1911 it gave up. It rescinded the hereditary status of all 79,300 *chōbu* of Yamanashi *seden goryō*. Later in the month, it transferred all (or almost all) of its holdings, a reported 298,237 *chōbu*, to Yamanashi authorities as *kan'yūrin*.[32] It was the single greatest failure in the history of Goryōrin and destroyed the vision of a Fuji-centered imperial foundation.

5

Managing Goryōrin: Plans and Performance

The Yamanashi experience was the only large-scale disaster in Goryōrin history. Elsewhere, woodland arrangements gradually worked out. Property Office personnel devised and implemented management plans, and while actual practice varied considerably from region to region, the forest as a whole did in fact develop into a reasonably-well-managed and profitable fiscal foundation for the Imperial Household.

AN OVERVIEW

"Management" of the imperial estate can be broadly or narrowly defined. In a broad sense it included the shift from a multi-asset to a more purely forest-based domain, the methods employed in dealing with competing interest groups, and the procedures used to dispose of unwanted parcels. Here, however, we use the term more narrowly to mean the handling of those parcels that were envisaged and treated as permanent possessions, whether or not formally designated *seden goryō*.

In this narrower sense, management pertained much less to the human-relations issues that were central to our earlier chapters than to issues involving the rest of the Goryōrin ecosystem, its woody-stemmed vegetation in particular. As the problems of human resistance and human entanglement gradually eased, the managers of Goryōrin focused more fully on their woodland biota, and our attention shifts accordingly.

The first step in managing Goryōrin was the development of general plans that defined the scope and content of policy for the foreseeable future. These general plans established guidelines for detailed surveying, mapping, and stand evaluation on all Household woodland. They also guided the preparation of local operating plans, which were to guide actual afforestation, aftercare, harvesting, and disposal of the yield at specific sites, as well as any construction projects undertaken to expedite those tasks. This Goryōrin management took clear shape from about

1897 onward, by which time the basic surveying and boundary marking were well underway.

Even where those preliminary tasks were completed, however, the managing of Goryōrin remained a difficult task for reasons of geography and history. They produced particularly clear dissimilarities in the management of Goryōrin in Hokkaido and Naichi.

Regional Differences: Hokkaido and Naichi

Imperial Forest in Hokkaido was mainly located in a half-dozen large parcels in the island's west-central section (see Map 7). It encompassed grand vistas of scarcely-worked woodland rich with old-growth stands of mixed conifers and broadleafs. Although mostly situated in mountainous terrain, it did not include the island's highest regions, and much consisted of low hills that opened rather gently onto riverine and coastal lowlands.

From the outset, Hokkaido Goryōrin was logged in an ad hoc "gradualist" (*zenbatsu*) manner, with logging contractors taking the most merchantable and most accessible trees from contracted parcels, advancing parcel by parcel ever deeper into virgin stands. In essence the method was classic exploitation forestry.

Almost from the start, however, Goryōrin managers made efforts to improve silvicultural practice on the island. They called on loggers to remove defective trees as well as good ones, to thin dense stands, care for young growth, leave desirable seed trees, and regulate the cut with goals of stand improvement and sustainable harvest at least partially in mind. Their efforts fared poorly, however, in part because the small staff at the Sapporo regional headquarters was ill-equipped to oversee its great acreage of virgin timber. More basically, the problems stemmed from difficulty of transport and scarcity of labor, which led lumbermen to log only if they could minimize their costs as far as possible.

Naichi Goryōrin, by contrast, was a maze of relatively small parcels. Most were situated on steeply sloping mountainsides that faced narrow valleys cut by small, rushing streams. The parcels were of maddeningly convoluted shape, their boundaries determined basically by the tortuous configurations of long-standing agricultural land usage, which reflected local topography. Even in the Kiso Valley, pride and joy of the system, Goryōrin was situated, as Map 5 suggests, in the upper reaches of watersheds and among peaks and ridgelines, while the flatter lands and lower hillsides along rivers were left in the hands of villagers. And despite attempts to consolidate or dispose of small parcels, many remained as scattered islets of Goryōrin in a sea of other holdings.

Furthermore, so much Naichi Goryōrin had been subjected to centuries of use that long before 1890 it already ranged from nearly denuded brush land through areas of mediocre mixed forest to densely spaced plantation stands of *sugi* and *hinoki*. With so many sites severely ill-wooded or already in even-aged monoculture stands, the Property Office from early on promoted – and to considerable extent practiced – clearcutting and replacement planting, a pattern quite unlike that in Hokkaido.

Fig. 5-1. Plantation forest in Kiso. Long-managed plantation forests, as in this early Meiji photograph of the Kiso Valley, were thinned and clearcut in rotation sequence. Here young seedlings grow among the decaying stumps of the former stand while the adjacent stand, already thinned, continues moving toward maturity.(Reproduced courtesy of Dean John C. Gordon, from a set of photographs in the Yale School of Forestry and Environmental Studies.)

In other ways, as well, history complicated management of Naichi Goryōrin. In particular the earlier-noted legacy of local use rights and the administrative compromises made by authorities during the 1870s-1880s encumbered much of it with an array of customary uses and rental agreements. As a result, no matter how intently the planners focused on trees, they could never fully ignore the human factor in their Naichi dominion. Rather, it continually hampered their managerial flexibility, even as it provided them with the brute manpower essential to their enterprise.

Coping: the Overall Picture

The difficulty of managing Goryōrin was evident in the Household Ministry's continual tinkering with its system of regional headquarters and the gradual elaboration of its central Property Office (or Bureau of the Imperial Forest, as it is more properly known after 1907). For two decades the Ministry repeatedly opened and closed headquarters, shifting parcels of woodland from one to another and not achieving general organizational stability until 1914, after much of the large-scale shrinkage in Goryōrin had been completed.[1] From then on the Office maintained four regional headquarters. One in Sapporo was responsible for Goryōrin in Hokkaido; one in Tokyo handled the Tōhoku and Kantō regions; one in Nagoya oversaw the Tōkai district; and one in Kiso managed the Kiso properties, most of which lay within Nagano Prefecture.[2] (For acreage figures, see Table E-3 in Appendix E.)

While fiddling with its regional arrangements, the Property Office also pushed ahead with other tasks, developing plans, deploying officials, expanding staff and infrastructure, and implementing its programs of afforestation, aftercare, and logging. The central staff of thirty-three regular officials in 1889 grew by 1910 to sixty-one regulars and fifty-three assistants, who oversaw additional personnel operating the regional headquarters and local offices. In 1946, when the Bureau of the Imperial Forest was about to be dissolved and the entire operation transferred to the Ministry of Agriculture and Forestry (Nōrinshō), it reportedly consisted of a central office, six regional headquarters, sixty-eight branch offices, and 299 section centers, the whole enterprise staffed by 1,829 administrative and professional personnel.[3]

In the end, then, Goryōrin staff constituted nearly a third of the Imperial Household's 6,000 total. And the Household's assets in stumpage, woodland, and forest infrastructure were valued at 2.6 billion yen (see Table 5-1 below). That sum was about 70% of its total assets, which included several palaces and villas and scores of tomb sites, as well as art treasures and paper assets (see Appendix B). [4]

As this trajectory of growth implies, from the time the Property Office acquired woodland to administer in 1889, it deployed officials and began implementing its programs, expanding operations and improving silvicultural practice as rapidly as it could. These trends are richly evident in the character and evolution of both management plans and managerial

Table 5-1: Assets of the Bureau of the Imperial Forest (Teishitsu rin'yakyoku), 1946 (in 1000 yen)

Category	*Location*	*Plantation Stands*	*Natural Stands*	*Total*
IMMOVABLE PROPERTY				
	Naichi	386,260	449,453	835,713
Stumpage	Hokkaido	38,445	764,596	803,041
	Subtotal	424,705	1,214,049	1,638,754
Land				635,574
Buildings				19,586
Immovable total				2,293,914
MOVABLE PROPERTY				
Forest Railroads				308,487
GRAND TOTAL				2,602,401

(Source: Hagino Toshio, *Nihon gendai rinsei no sengō katei*, p. 66, Table 9.)

performance, which we treat below, and in the several aspects of Goryōrin operations that we examine in Chapter 6.

MANAGEMENT PLANS

During its lifetime, the Property Office developed three general management plans: the first in 1898–99, a slightly modified one in 1908–10, and an ambitious "Hundred Year Plan" first drafted in 1927 and reworked into final form by 1934.[5] Each plan set a numerical goal for the amount of Goryōrin acreage that was to be carefully surveyed, mapped, and brought under complete management within the plan's lifetime. It specified what the management of that acreage was to entail and spelled out budget allocations for the plan's several parts.

Early Plans

The first plan distinguished three basic categories of land. *Ringyōgaichi*, or "non-timber land," denoted sites that would never have value for timber production, notably mountain peaks, cliffs, swamps, and arable. Those areas could be retained for their aesthetic or other value or disposed of as seemed fitting. *Jun ringyōchi*, or "potential timberland," was acreage that lacked satisfactory stands but could be developed into productive timber forest (see Fig. 3-1). *Ringyōchi*, or "productive timberland," denoted the most valued areas, those with appropriately dense stands of desirable trees, and it was the category to which forest planners aspired to elevate all *jun ringyōchi* (see Fig. 5-1).

To implement the policies of improvement that would gradually transform potential into productive timberland, the several regional headquarters were to organize the woodland they managed into operations areas (*jigyōku*) and to prepare site-specific management plans for each area.

These operations areas were to range in size from about 1,000 to 10,000 *chōbu* or more. Each was to be subdivided into rotation blocks (*rinbatsuku*), and each block would consist of parcels so organized as to be processed in sequence for rotation felling. In accordance with the condition of each parcel and the species growing there, programs of rotation felling, planting, and aftercare were to be implemented so as to ensure that each rotation block developed into an ever-productive forest unit, with harvestable trees maturing in sequence on parcel after parcel, doing so for evermore.

The rotation periods, from planting through thinning cuts to the final cut, were long, ranging from 60 to 120 years or even more, depending on tree species and site condition. (For the cutting schedule of the Hundred Year Plan, see Table E-4 in Appendix E.) We noted in Chapter 3, in an early example from Gunma Prefecture, that the Property Office restricted rental-land harvest cycles for timber production to thirty years and for other uses to shorter periods, durations that were adequate for most of a farmer's needs (see Fig. D-1, Appendix D). For its own logging operations, however, it specified the much longer cycles that would yield greater volumes of large-size construction timber.

In a basic sense, the Property Office adhered to the policies outlined in its two early plans for nearly as long as Goryōrin survived. As we note below, however, the particulars of actual practice deviated from such plans, especially in Hokkaido. Moreover, the Office modified specific aspects of policy as problems were discovered and new silvicultural knowledge was acquired.

In particular, although its basic goal of developing stands of fully mature trees by setting life-cycle goals of 60–120 years endured, the elaborate rotation-block scheme of the early plans proved unworkable. In Naichi it failed mainly because of Goryōrin's steeply mountainous character, geographic irregularity, and stand diversity, and in Hokkaido because of the forest's old-growth character, transport problems, and labor scarcity. The scheme also lost appeal as new silvicultural theories gained favor.

Surely, too, the closely regulated rotation-block scheme was undermined by the Household Ministry's responses to sporadic national emergencies, in which ministers called for intensified logging even at the expense of management plans. This occurred during the Russo-Japanese War of 1904-05, after an earthquake ravaged the Tokyo-Yokohama region in 1923, and during and after the Pacific War. In any case, by late Meiji the rotation-block system was being questioned, and during the 1920s it nearly disappeared from both plan and practice.

The Hundred Year Plan

The ambitious Hundred Year Plan deserves attention because of the gains in silvicultural sophistication that it reveals and because of what it shows about long-term objectives.

As a whole the Plan addressed four topics. On forest management proper it adhered in general to policies as they had taken shape during the twenties, with detailed sections discussing means of boundary delineation and policies for afforestation, aftercare, harvesting, and the handling of road construction and other forest engineering projects. It also dealt with the technical tasks of mensuration: calculating stumpage, the harvest, work-loads, sustainable harvest rates, necessary reforestation rates, and the fiscal, spatial, and material requirements of construction projects. Thirdly, it detailed personnel needs in terms of administrative and scientific skills, as well as particulars of organizational structure and research and operations facilities. And finally, it drew these matters together in a detailed estimate of income and expenditure.

The planners envisaged a future of improvement rather than growth. Eventually, after disposing of the remaining unwanted parcels, they aimed to stabilize Goryōrin acreage at 1,275,000 *chōbu*. All but 175,000 of that would qualify as *ringyōchi*. They expected to continue making both selective and clear cuts, but strongly favored the former, in line with the newest silvicultural thought of the day. Indeed, they would restrict clearcutting to about four percent of harvested acreage.[6] They would employ rotation schedules as seemed apt, adapting German yield tables to guide those schedules so as to produce a slow increase in the volume of the annual cut.[7] The planners aimed to accelerate planting on unstocked parcels and scheduled more frequent improvement work on areas of selective felling.

They also planned basic changes in harvest procedure. Hitherto, they had relied heavily on entrepreneurial contractors for logging and log removal, selling them stumpage, logs on the ground and in storage, and fire-damaged stands for reduction to charcoal. In future, they aimed to handle more and more of their own processing, within a century doing all of their own logging and half the charcoal work. Sawmill work, however, they would leave to entrepreneurs, having found it uneconomic to operate their own mills, as we note in Chapter 6.

Infrastructure work also was to continue. Planners intended to have laid out within a century 2,943 km of forest railroads and tramways, 2,696 km of wood roads, and some 36,000 km of work paths. By then, too, they planned to have set up ninety-four branch offices, each staffed by about six professionals, operating under the regional headquarters. And Goryōrin would be further sub-divided into 571 sections, each with a local office where foresters could provide close oversight for their operations areas.

The rationale for this ambitious scheme reflected the changing times. As a senior figure explained, the new plan was needed because in future, unlike the past, forest must meet not just local needs but national and international demands. And not just for construction and fuel wood, but for many industrial purposes. Planning for those needs would be difficult, so the Property Office required this new plan to improve long-term forest vitality and utility.[8] The Hundred Year Plan was thus ambitious, a grand vision, one meant to inspire generations of foresters to dedicated service as they confronted a rapidly-changing world.

MANAGERIAL PERFORMANCE

Such, in brief, were Property Office plans. As every day reminds us, however, plans are one thing; performance, another. Accessible evidence lets us examine, in overall terms, the performance of Property Office foresters generally, and, in more detail, the Sanru operations area in Hokkaido.

Overall

The Property Office's three general plans closely paralleled trends in Forest Agency policy on *kokuyūrin.* The plans of both Office and Agency addressed the same basic portfolio of problems, all doing so with relatively long-term fiscal purposes in mind. But the particulars of policy content and outcome changed as decades passed.

1. *Early Plans*

The general plans of 1899 and 1910 called for development of local operational plans tailored to specific operations areas. And under the guidance of regional headquarters, foresters did prepare them. By 1906, such local plans existed, at least in preliminary form, for eighty-seven operations areas. They covered a registered 509,163 *chōbu,* roughly a quarter of the total Goryōrin.[9] In following years the staffs at regional headquarters continued developing such plans and by the twenties had completed them for nearly all their holdings.

Like general plans, however, these local ones were only guidelines, and drawing them up did not assure their implementation. Overall one can say that from late Meiji onward the Property Office gradually attained some of its objectives, but not all. The initial tasks – recruiting and training staff; establishing branch and local offices; estimating acreage; evaluating stand composition, quality, and volume; demarcating operations areas and rotation blocks; categorizing sites as *ringyōchi,* etc.; drafting, revising, and getting formal approval of local plans; and the actual start of culling and improvement – all made good progress, at least in quantitative terms and particularly in Naichi.

Thus, the first plan aimed to encompass 324,000 *chōbu* by 1907, but on completion that year had actually covered 364,000 in Naichi and another 40,000 in Hokkaido. The second plan aimed to cover 300,000 more in Naichi and 700,000 in Hokkaido. But by 1928, when finally completed, it had processed a total of 1,110,000 *chōbu* in 162 operations areas. And areal surveys had by then recorded a Goryōrin total of 1,517,000 *chōbu.*[10] The accuracy of mensuration was still unsatisfactory, however, especially in Hokkaido, so these figures, and even more so estimates of stand volume, continued to be treated cautiously.

Table E-5, which gives slightly different figures, reports management results at ten-year intervals. It shows, until the later 1920s, a decline in the proportion of Goryōrin that qualified as *ringyōchi.* This trend existed because the acreage defined as "Managed Area," meaning areas for which operational plans had been finalized, kept growing as surveyors completed their assessments of more and more parcels. That process identified additional *jun ringyōchi* faster than workers could upgrade sites to *ringyōchi* status, creating the statistical appearance of a deteriorating woodland.

By 1938, most Goryōrin (1.3 million *chōbu*) had been assessed. And the actual acreage defined as *ringyōchi* had nearly quadrupled since 1908, reaching 1.09 million *chōbu*. Much of that increase, however, was in Hokkaido, and much simply constituted a reclassification as *ringyōchi* of old-growth stands of mixed vegetation that were supposed to be selectively cut and gradually improved. So a significant portion of the Household's "productive timberland" was neither plantation stands amenable to clearcutting and hand-planted restocking nor even the vibrantly healthy, mixed, multi-aged stands of well-managed forest that theory of the day envisioned. Rather, it was old-growth woodland that met the local plan's minimal stand-requirement per *chōbu*.

2. *The Hundred Year Plan*

For all its grand ambition, in the end, the Hundred Year Plan fared even less well than the early ones. But it started out with great promise in September 1936 when the Property Office issued directives for its implementation. These permitted branch offices to exercise latitude on specifics, thus acknowledging the demands of geographic diversity while also enforcing the overall thrust of policy. And initially those offices experienced good progress in implementing the grand new plan.

Indeed, because the planning had been in progress for years, by 1936 some regional headquarters were already developing operational plans that followed the new guidelines. They opened and staffed more branch offices and section centers and put into use newer technologies, most notably that for hauling timber out of the forest. They adopted newer yield tables and mensuration techniques and otherwise worked toward more effective and efficient operation and long-range forest enhancement.

With the coming of war in 1937, however, trends started to reverse. From 1938 onward, gas rationing caused gasoline-powered equipment which had begun to be used in logging during the preceding decade to lie idle. Forced back to older methods of timber transport, the logging work became more labor intensive. Nevertheless, the Property Office gradually increased its rate of extraction, doing so in part by reducing hand-planted reforestation and aftercare. This was so especially from 1940 onward, as growing military demand on the one hand and scarcities of labor and matériel on the other altered work priorities.

As the battlefield situation deteriorated, military control of provisioning tightened, bringing Goryōrin, *kokuyūrin*, and all categories of *min'yūrin* under closer direction by prefectural organs responsive to military command. However, repeated attempts to clarify and coordinate chains of command produced about as much infighting as efficiency among landowners' and lumbermen's guilds, civil government organs, and military authorities.

In April 1943, after major naval reverses in the Coral Sea, Solomon Islands, and elsewhere had created frantic demands for more wood with which to build more warships and planes, the Hundred Year Plan was formally suspended. Then the Property Office notified regional headquarters that they must commence "special harvesting" work to aid the war effort and drafted special rules to guide the accelerated work. In essence these rules licensed a de facto reversion to classic exploitation forestry: the fullest possible logging of wanted types of wood from the most accessible and convenient places.

The results were soon evident. Despite labor shortages and transport difficulties, by one calculation, the 1942 harvest of 8.6 million *koku*, which was already well above that of prior years, rose to 10.7 million in 1943 and 10.97 in 1944.[11] (For a fuller set of figures, see Table E-11.) And much of the increase was achieved by government-organized logging crews: in 1943, their peak year, such crews produced 94.8% of the 2,698,000 *koku* of logs taken from Naichi Goryōrin.[12]

After the war ended, intensified logging continued as the country rebuilt its war-ruined urban infrastructure, and by the time that trend was reversed, Goryōrin had ceased to exist. In the final years of Imperial Forest, then, exigencies of the day had rendered management plans moot, and the resulting woodland depredation not only created major challenges for the next generation of Japan's foresters but also concealed a record of managerial planning and practice that had been reasonably successful.

The Sanru Case

Scholarly studies of the Sanru operations area afford us a closer look at actual performance on Goryōrin.[13]

In Hokkaido, as noted earlier, planning was implemented more slowly than in Naichi, and a local operational plan for the northerly Sanru area was not developed until 1920–21. Officials there completed the first careful survey of their acreage and stand in 1920, declaring that it consisted of 17,000 *chōbu* of *ringyōchi* and 18,000 of *ringyōgaichi* (over 6,000 in agricultural leaseholds, as noted earlier). On the basis of that analysis, and following the guidelines of the second general plan, their local plan divided the 17,000 *chōbu* of productive timberland into four rotation blocks, each with four or five parcels. The plan called for a 120-year life cycle, with each parcel to be revisited for thinning and final cuts at forty-year intervals. The loggers, who were permitted to cut four parcels every decade, were authorized to extract about 1,000,000 *koku* of timber per cut, or roughly 33–50 percent of existing stumpage, it being estimated that the forest's growth rate would make that a fully sustainable level of harvest.[14] The cut was to be made, moreover, so as to enhance natural regeneration.[15]

For a decade Sanru officials followed these guidelines. Then a ten-year assessment in 1931 provided the information they needed to develop a new operational plan that reflected Tokyo's emerging Hundred Year Plan.

Employing the metric system – itself a sign of the changing times – the new ten-year Sanru plan organized 19,600 hectares (99.5% *ringyōchi*) of *todomatsu* (todomatsu: *Abies* sp. Sakhalin fir), *ezomatsu* (ezomatsu: *Picea jezoensis*), and select broadleafs in four rotation blocks to be selectively logged at thirty-year intervals in a 120-year life cycle. A less well-wooded 1,800 hectares were to be organized as two small rotation blocks of *todomatsu* and *ezomatsu* for clearcutting and restocking in 100-year cycles. Foresters estimated that the large former area would yield 380,000 cubic meters (about 1,407,000 *koku*; 19.4 m^3/ha) of wood by 1941, and the latter 31,000 (17.2 m^3/ha).

The logging was to be linked to stand improvement and restocking. Natural regeneration was to be promoted on the selectively-cut parcels. And the assistance to seedlings was to include clearance of underbrush and trimming of vines for up to seven years. Also, depending on circumstances, workmen were to pursue supplemental planting of *todomatsu* seedlings at a rate of 300–1,000 per hectare, but not to exceed 20% of a fully stocked site's seedling count. The areas for clearcutting and hand-planting were to be logged as the smallest possible parcels, with care to protect existing young growth. On these sites, planting projects were to be less than 5–10 hectares in size and were to aim at mixed stands with a minimum of 20% broadleaf content. Finally, to expedite operations, the plan called for construction of 18 km of mainline forest railway, 6 km of branch line, 9.5 km of tramway, and 31 km of wood road.

Regarding actual performance in Sanru during the decade to 1941, construction of transport facilities made good progress, although the planned branch rail lines were not completed and 3.6 km of planned wood road remained unopened. In logging work, the overall felling rate lagged a bit, standing at about 35% of stumpage rather than the intended 40%. In selective-cutting areas, harvests accorded reasonably well with the plan. And enough self-seeded new growth survived so that supplemental planting rates there were generally lower than planners had anticipated, though they varied greatly from site to site. On the clearcut areas matters went a bit less well. Although larger than intended, the sites yielded only about three-quarters of the estimated cut. And foresters ended up hand-planting 277 hectares, although they had anticipated planting only 245.

During the thirties, Sanru officialdom continued to rely heavily on local farmers for both felling and planting work. And as gasoline-powered equipment was lying idle and wartime demand for timber intensified, the burden on manual labor increased.

During the early forties, steam engines could still be used, so they kept operating, and another six kilometers of forest railway were added, along with three kilometres of wood road. In response to wartime pressure, however, the demand for timber increased so much that Sanru loggers took smaller, dead, and diseased trees, made felling decisions on the basis of convenience rather than forest well-being, and gave little attention to shielding replacement stock from damage.

As for restocking, with so many men off to war, students and diverse others were employed in such planting work as continued to be done. Nevertheless, the rate of replanting dropped sharply due to shortages of trained labor for maintaining seedling beds, clearing areas to plant, and providing aftercare. And the success rate in transplanting, which generally had not been very high anyway, deteriorated, with most seedlings withering or being shaded out by uncontrolled overgrowth.

In the end, then, the Sanru version of the Hundred Year Plan fared poorly. But for nearly a decade it had enjoyed generally successful implementation, even in the context of World Depression, political turmoil, and ultimately disastrous war. As in Goryōrin generally, the bigger picture that emerges from Sanru is one of increasingly ambitious planning that was accompanied by improving performance until broader political factors confounded planners and performers alike during the later 1930s-1940s. These trends show up more clearly, perhaps, in the next chapter, where we look more closely at select aspects of forest practice.

6

Managing Goryōrin: Aspects of Practice

In the end, history bypassed Goryōrin, its planners, and plans. By then, however, the professional foresters and work crews operating under Property Office authority had made a lasting impact on woodland of the realm. The several aspects of their effort were deeply interconnected, of course, and disaggregating them is awkward, but doing so may help clarify what they did and did not achieve, why, and with what consequences. To that end, their exertions can be examined in terms of forest mensuration and research, afforestation and aftercare, technological adaptation, logging practice, and, by way of summation, a look at the outcome.

MENSURATION AND RESEARCH

Forest mensuration, one of the world's more arcane arts, commands attention because it was central to the Property Office's entire managerial effort. The relatively simple task of measuring acreage gradually gained accuracy as surveyors were trained and deployed and as surveyor's levels and the techniques of triangulation were introduced and adopted. The much more difficult tasks of estimating stumpage volume and predicting timber yield advanced more slowly. But "eyeballing" did give way to increasingly refined techniques of estimation. And programs of research gradually produced yield tables appropriate to particular locations and tree species throughout Japan.[1]

To elaborate, initially Goryōrin foresters employed the simple measuring system adopted by the Forest Agency in its 1879 mensuration manual, *Shakujime shidashi benran*. It applied a single eyeballing technique to all species of tree, instructing foresters to measure a tree's diameter at eye level and to figure a one *sun* (3.03 cm) reduction in diameter for every *ken* (1.82 m) of trunk height. The method was useful for measuring conifers growing in dense plantation stands, but it seriously overstated trunk volume in most natural stands. So the guideline was modified a few years later by the simple expedient of applying it to only fifty percent of the stumpage on such sites.

In its management plan of 1899, the Property Office spelled out new methods of mensuration. These called, in some circumstances, for the measure of every tree; in others, for the establishment of sample plots from which general estimates could be made. In the latter case stumpage on a sample plot was to be measured in this five-step manner:

1. Measure the diameter of each tree [in the sample plot] at eye level (4 *shaku* 3 *sun* above the ground) using an outside caliper (*rinjaku*). But round up any fraction over a half-*sun*. [1 shaku = 30.3 cm = 11.93 inches; 1 sun = one-tenth shaku]
2. Classify all the trees [by size] into five or fewer classes and calculate the total basal area [at eye level] for each class.
3. Using the basal area of each class, estimate the average eye-level diameter and use this figure as the sample-tree diameter [for this class].
4. Selecting a sample tree in each class, have it felled and measure its volume of wood [using the mid-diameter measure of each 2-*ken* log].
5. By comparing the basal area of the sample tree to that of the entire class, calculate wood volume for the class and thence for the whole sample plot.

In later years the Property Office tightened its mensuration guidelines, but it continued to advocate both full stem counts and estimates based on sample plots.

Thus regulations of 1913 specified that in preparing stumpage estimates for parcels up to 10 *chōbu* in size, one should select at least two sample plots, each one one-twentieth to one-fiftieth the parcel's size; for parcels over 10 *chōbu,* use three sample plots one-fiftieth to one-hundredth the parcel's size. For estimating sample plot stumpage, they replaced the Tokugawa-era "eye-level" method with "diameter at breast height" (dbh), a formulation that in fact changed the height of the measurement very little while seeming better suited to workers using calipers. And whereas the procedure of 1899 called for creation of five or fewer tree-size classes when determining sample trees, that of 1913 stipulated a separate class for each *sun* of increase in diameter. Lastly, it required a new, reputedly more accurate method for determining volume of the sample tree in each class.

Household foresters also started developing their own yield tables to provide estimates of the volume of wood in plantation stands of select species in designated areas and to permit projections of growth so that they could calculate sustainable rates of harvest. In practice, however, errors of mensuration and flaws in yield tables contributed to actual harvests that were well over or under estimates, to the dismay of either the Property Office or the logging contractor. By the 1920s, this situation, along with general development of the profession, was prompting active forest research. During the thirties, as the planted acreage of Goryōrin expanded, foresters were able to develop more reliable yield tables, such as one prepared in 1936 for *hinoki* (*Chamaecyparis obtusa*; Japanese cypress) growing in the Adera River operations area in southern Kiso or another for *todomatsu* in the operations area southeast of Otaru in Hokkaido.[2]

Besides research to improve harvest calculations, Household foresters undertook studies of specimen growth, soil acidity and its effects, factors affecting seed-germination rates, and other projects designed to improve forest productivity. And they engaged in vigorous argumentation about policy and procedures, their views showing up in such professional journals as *Goryōrin* and *Sanrin* and in the publications of lumbermen's and forest owners' guilds. From a later vantage point this legacy of research and mensuration theory and practice may seem rather primitive, but compared to the forest practice of early Meiji, it was appreciably improved. And that improvement contributed to better managerial performance in Goryōrin.

AFFORESTATION AND AFTERCARE

A key purpose of forest research was to enhance the results of afforestation and aftercare. Until about 1940, the main story was one of changing and improving technique and expanding scale, in Naichi initially and in Hokkaido from about World War I onward.[3]

Naichi

Goryōrin foresters inherited from their Tokugawa forebears a rich legacy of silvicultural advice on tree-planting and nurturing. And they added more from Europe as the years passed.[4]

During the 1890s, the Tokugawa legacy was particularly evident in the techniques employed on Naichi sites, where Household planters reforested about 1,000 *chōbu* per year on average. Some was natural restocking: allowing or assisting seed from on-site trees to sprout and grow. More, however, was hand-planting of seedlings, with some sets taken from seedbeds, others from self-seeded sites. Such seedling work might entail a complete replanting of clearcut parcels or supplemental planting to fill gaps on naturally restocked locations. Planting technique depended on the tree species and planter's expertise, with density on clearcut sites ranging widely from 1,000 to 10,000 seedlings per *chōbu*. *Sugi* (cryptomeria; *Cryptomeria Japonica*) was a major species, but throughout the decade a grand variety of other evergreen and deciduous species was also planted.

The general management plan of 1899 promoted standardization by defining afforestation policy more clearly. It called for planting all logged areas within five years of their clearance and all treeless sites within twenty years. Tree-planters were to set out 3,000 to 6,000 seedlings per *chōbu*. It also recommended natural reseeding and advised the mixing of two or more species on each site. The plan also called for aftercare – to clear competing growth regularly and to thin growing young stands frequently.

At the time, planting was almost exclusively a Naichi enterprise, and naturally reseeded areas evidently tended to be neglected. So, in the second plan, in 1910, emphasis shifted to hand planting even though it was immensely more costly than unsupplemented natural restocking.[5] That plan also speeded up the replanting of logged sites, saying it should be done within three years rather than five. In practice, the acreage being hand-planted each year gradually expanded, rising to around 3,000 *chōbu* by 1910, to 4,500 by the end of World War I, and exceeding 5,000 per year in the mid-thirties, before its precipitous wartime decline (see Tables E-8 and E-9).

As the decades passed, planters increasingly favored stocking with *hinoki* in Naichi. It matures more slowly than *sugi* but produces lumber with an appealing scent and finer cellular structure, making it more attractive and durable. In practice, whereas less than eight percent of seedlings had been *hinoki* during the 1890s, by 1920 the figure was approaching sixty percent. By then much of the species diversity that marked early planting had disappeared, and seedling density was slightly reduced to 3,000–5,000 per *chōbu*. Also, planters were instructed to provide better aftercare than hitherto.

By the 1920s, woodsmen were being introduced to the principles of plant succession and continuous forest.[6] These concepts renewed their interest in more diverse stands, natural reseeding, the mixing of natural and hand-planting, and further reduction of seedling density to 3,000–4,500 per *chōbu*. Planting work was facilitated by the establishment of more seedbeds, which by then were enabling planters to set out three-to-four year old seedlings rather than the more vulnerable one-to-two year olds of earlier years. Also, the increasing mileage of railroads and wood roads expedited deployment of planters and seedlings alike. And in following years these trends continued, until wartime labor shortages forced curtailment of planting.

Hokkaido

By the 1920s planting had also become established as a part of forest practice in Hokkaido Goryōrin.[7] The burgeoning Naichi economy had by 1910 created a demand for pulp and saw timber that exceeded the region's available resources. So entrepreneurial mill operators turned to Hokkaido to meet their needs, erecting mills on the island and starting to consume its old-growth stands. The "gradualist" logging strategy that they utilized was by World War I producing an ever-larger acreage that warranted restocking because it was relatively accessible and, therefore, more profitable to nurture than the deeper forests into which felling crews were advancing. Accordingly, officials despatched by the Sapporo regional headquarters hired people, mainly local farmers, to undertake small supplemental planting projects. They favored local spruce (*ezomatsu* and *todomatsu*) in particular, but also planted select broadleafs (notably the ash, *yachidamo: Fraxinus mandshurica*). And as years passed, they set out a growing proportion of the recently introduced exotic, German spruce (*Doitsu tōhi: Picea abies*).

As in Naichi, much of the early supplemental planting actually failed, in considerable part for want of adequate aftercare. These poor results were discouraging, and during the early twenties a severe recession in the lumber industry (the bust phase of the World War I boom-and-bust cycle) undercut enthusiasm for reforestation. By the decade's end, however, the lumber market was beginning to recover, the Hundred Year Plan was taking shape, and national interest in providing financial assistance to hard-pressed farmers was becoming more pronounced. The Sapporo headquarters (and the one in Asahikawa after its establishment in 1928) responded to these trends by developing more ambitious programs of full restocking on clearcut parcels and supplemental planting on selectively felled sites, hiring locals to implement them.

During the thirties, afforestation work in Hokkaido Goryōrin intensified, primarily as supplemental planting in selectively cut areas. But closer attention was given to procedure in hopes of improving survival rates. Household foresters grew some seedlings and purchased others, and they hired crews (*kumi*) of four-to-five workers to do the planting. Many of the crew chiefs (*kumi gashira*) were professionals brought in from Naichi because of their reputed ability to get better results. The crews (mostly farmers, but more and more students during the war years) were instructed by their chief to prepare their site carefully and plant at the prescribed rates of 350–1,000 seedlings per *chōbu* on supplemental projects and about 3,000 per *chōbu* on clearcut sites. The latter work cost about five-to-ten times that of the former.

In the Sanru operations area, seedlings came from temporary local beds until permanent seedbeds were established in 1938. Depending on the species, the sprouts were nurtured for two to five years before being set out. Planting instructions called for work crews to prepare patches of soil ten to twenty meters square or larger, such openings being sufficient to assure seeds and sets adequate sunlight among surviving natural stock. They were to prepare the patches by raking them to rid the soil of troublesome roots and to smooth its surface. They also were to clear away weeds, brush, and obstructions and cut nearby vines. The aftercare tasks of brush and vine clearance, follow-up planting, and thinning of self-seeded sites were to continue for six years.

Practice, it appears, generally followed regional office guidelines. But because labor shortages eventually compromised the quality of both planting and aftercare and because seedling stock was insufficient for all the recently logged areas, the results were disappointing. Some stands thrived, but on many sites work remained for a postwar generation to redo and complete.

Reforestation is by its nature a tricky undertaking, and the overall record on Goryōrin was decidedly mixed. Until the wartime years, however, the scale of planting gradually increased and the quality improved, especially in Naichi. In consequence, by 1946, as Table 5-1 in Chapter 5 indicates, over forty-five percent of the estimated asset value of stumpage in Naichi Imperial Forest was in its plantation stands, as was nearly five percent of that in Hokkaido.

CHANGES IN TECHNOLOGY.

Both forest research and afforestation were intended primarily to increase the long-term yield of logging. The same goal lay behind technological changes that contributed to the long-term expansion in Goryōrin output. Three changes in particular – in methods of felling, transporting, and milling – merit attention.

Crosscut saws

The least well documented of these changes in technology was the simplest, the introduction of crosscut saws for felling large trees. They had been used in Japan during the sixteenth century, but during the Tokugawa period their use had been proscribed to discourage timber theft. Fellers had used axes instead, but after 1868 such saws became a standard logger's tool. Their use expedited the harvest, especially of very large trees. Previously those had simply been left standing as too big to handle, or they had been felled by a time-consuming process of chopping and burning away at the base until the tree toppled over. (The chain saw, the next major change in felling technology, was not introduced to Japan until 1947.)

Roads and Railroads

Far more noticeable to observers than the adoption of crosscut saws were changes in transport technique.

Waterways had long been the key arteries for moving timber, but by 1885, as knowledge of European methods of timber transport became available, discussion of their applicability to Japan began appearing.[8] Commentators pointed out the severe disadvantages of reliance on rivers – their seasonality, unpredictability, and slowness, and the tendency of ledgy passages to damage timber. And they examined the issues of railroad construction costs and the utility of vehicular roads in Japan's steeply mountainous terrain.

Talk was one thing, however; action, another. The chief obstacle turned out to be social rather than economic. But in due course it was overcome, as we can see by looking at the experience of Goryōrin in Kiso and Hokkaido.

1. *Overcoming Resistance*

Even when the economic arguments clearly favored use of road and railroad, social considerations spoke otherwise. In Naichi, after all, the existing timber transport system constituted a major set of vested interests.

Fig. 6-1. Log Transport in Kiso. Changes in timber processing proceeded gradually. In this early Meiji photograph, logs are still being floated down the Kiso River, but they are being felled by crosscut saw, not axe. And they are being sent downstream as logs, not as hewn timbers, to be sawn by mechanical saw rather than handsaw. (Reproduced courtesy of Dean John C. Gordon, from a set of photographs in the Yale School of Forestry and Environmental Studies.)

Professionals skilled in the several stages of the work – from felling to skidding to chuting to free floating to rafting to beaching and stacking at storage sites – were organized in guilds, proud of their art, and solicitous of their interests. And these customary arrangements and interests were embedded in a rhetoric of regional expertise and excellence that had long been central to the employment of forest workers and the marketing of their products.

To suggest the strength of this rhetoric, when lumber companies sought river bosses to manage projects in Hokkaido, they commonly hired experts from Kiso and Kumano, regions with storied legacies of damming, chuting, free floating, and rafting. When Sapporo officials decided to handle some logging themselves during the 1920s, they hired crew chiefs from the Kiso Goryōrin to supervise the actual felling work.[9] In Naichi, *sugi* seedlings from Yoshino, celebrated since the eighteenth century, continued to enjoy special repute and a broad market. And timber produced in the "Kiso style" or the "Yoshino style" or select products from Akita, the Tenryū valley, or wherever, commanded good prices because of their distinctive reputations. Proposed changes in techniques of felling and transport threatened public marketing images such as these and the social and material interests that stood behind them. Brand names had value, and people knew it.

So during the later 1880s-1890s and into the early twentieth century, desultory debates continued on the relative merits of "tradition" and "change" in timber transport. By the early 1900s, however, railroad mileage was multiplying. As the lumber market blossomed in response to a nationwide construction boom, the immense advantages of rail transport were proving irresistible. That trend gave rise to more urgent and heated debates about the pros and cons of the new technology, and resistance notwithstanding, the transport system was gradually transformed.

As of 1906, most Goryōrin timber, like that of other provenance, still went to market in the Tokugawa manner, via river and raft.[10] But as market demand had surged in preceding years and more accessible hillsides had been stripped of timber, the need for better forest entrée became compelling. In February of that year the Property Office established a Construction Agency (Dobokuka) with responsibility for building wood roads, tramways, forest railroads, and erosion-control works.

All of these projects would expedite logging, but the railroad, both within the forest and between forest and market, was the crucial factor. Compared to a river, it could haul so much more at so much greater speed and so much less cost, and do so with minimal regard for season. In following years, construction pressed ahead, especially in Kiso and Hokkaido, and by 1937 the Property Office could boast 343 km of forest railroad, 340 km of tramways, 667 km of wood roads, and 16,715 km of trails for use by work crews.[11] By then, moreover, nearly all the wood that these devices moved to the forest edge made its way from there to market via railroad and public highway.

2. *The Kiso Case*

The impact of railroads was evident in Kiso. Back in 1889, when the Household Ministry had acquired a recorded 399,786 *chōbu* of Kiso woodland in 532 parcels of former *kanrin*, it also claimed their downriver facilities. These included the oceanside lumberyards at Kuwana and Shiratori on Ise Bay and the log boom, riverside landing, and storage site at Nishikori in Gifu Prefecture, which lay nearly halfway from Kiso to the sea (see Map 4). A year later the Ministry designated Nishikori a hereditary possession and acquired enough nearby acreage to give it 44 *chōbu* of land for timber storage, administrative offices, and work-crew quarters.

For over a decade, Goryōrin timber continued to leave Kiso via Nishikori, but gradually changes began to appear. By 1900, construction work on the western end of the Chūō Line – destined to link Nagoya and Tokyo via the Kiso River valley – was approaching the lower Kiso. The following year the Kiso regional headquarters started building a short, lightweight forest railway along the Adera River, a precipitous, ledgy tributary of the Kiso, eventually extending the line ten kilometers up the valley. Built to expedite logging by hauling men and materials up the valley and logs back down, initially it just delivered logs to the Kiso for regular floating down to Nishikori.

By 1905, however, Chūō Line construction was entering Kiso from both north and south. As noted in Chapter 4, the line from Tokyo was in operation to Kōfu by 1904, and from there it pushed rapidly up to Suwa and Shiojiri, and then down the Kiso. By 1911 the two ends had been linked.

From about 1905 onward, as this Chūō Line construction pressed ahead, loggers in the valley revived discussions about the relative merits of rail and river. The issues at stake were re-argued, but that the new technology would prevail in Kiso was clearly implied in early 1906 in the Property Office's creation of its Construction Agency. In ensuing years, Household loggers sent more and more sticks down their forest railway to log yards adjacent the new Chūō line, from there sending ever more of them down to Nagoya by rail.

The opening of the completed Chūō Line in 1911 spurred forest railway construction in other branch valleys of the Kiso. Work on a line along the Ogawa River began in 1913 and was completed by 1916, its length 19.4 km. On the much longer Ōtaki River, work began in 1917 and continued into the 1930s. For Kiso Goryōrin as a whole there were by 1925 some 100 km of forest railway; and by 1937, 179 km. In 1928, mechanization reached deeper into the forest when Kiso loggers began using gasoline-powered skidders to snake logs out to assembly points. And, in 1940, they began using trucks to haul wood from more distant assembly points to railhead or sawmill.[12]

3. *Hokkaido*

The government promoted railroad construction in Hokkaido as a national defense policy, a part of its broader program for countering

Russian expansion by developing the island and binding it to Naichi. By 1903, a rail line from Sapporo had reached northward through Asahigawa to Nayoru, and in following years more and more lines spread across the island, bringing rail access to all major parcels of Goryōrin (see Map 7).

As this rail network developed, it displaced rafting operations on the Teshio and other rivers and helped turn the minor fishing village of Otaru, west of Sapporo, into a bustling timber port. Working for Mitsui Bussan and other lumber companies, Otaru's tens of thousands of laborers (76,900 at its peak in 1928) handled logs, lumber, and pulpwood brought by rail from forests throughout the island, whether those of government, Household, or entrepreneur. In addition, workmen at Otaru processed timber brought from Karafuto and Siberia for eventual re-shipment to Naichi or abroad.[13]

As in Kiso, so too in Hokkaido, growth of a rail network spurred changes in transport within the forests themselves. In the Household's Sanru operations area loggers started using imported five-ton Caterpillar tractors in 1929. Each could pull three or four sleds laden with 100 *koku* of logs, making two ten-mile round trips daily and moving some 14–15,000 *koku* per logging season.[14]

That transport capacity sharply increased the rate at which logs could be hauled from distant assembly sites to railhead or streamside. This increase overwhelmed the upper Teshio's flotation capacity, spurring the regional headquarters in 1931 to start constructing its planned Sanru forest railway. Trains started running on it in 1936, with two eight-ton, coal- and wood-burning steam engines pulling short strings of log cars that soon were carrying some 50,000 *koku* of logs per logging season. Laborers working with hand tools still moved most of that wood to assembly points, but there too mechanization made advances. More and more steam engines and Caterpillars worked in other parts of the Sanru and nearby operations areas, and more gasoline-powered engines were used on tramways.

In both Naichi and Hokkaido, then, new transport technology displaced old, changing the way wood was moved and permitting a sharp increase in the rate of extraction. As transport changed, moreover, so did the logging profession, particularly in Naichi. There the old guild arrangements fell apart. Fields of specialization were reconfigured; seasonal work patterns were ruptured, and peripatetic loggers, who took their new skills from job to job, became more common.

In Kiso, more narrowly, even as the volume and value of its timber output grew, use of the river declined. By 1920, the Property Office retained only 22 *chōbu* of land at Nishikori, and in the following year it removed the site from its hereditary holdings. By 1926 the last logs had left Kiso by river, and in 1931 the Office disposed of its property at Nishikori. Thus ended one of the most celebrated institutions in Japan's forest history.[15]

Sawmills

The other major technological change of these decades was the introduction of mills – for saw timber in 1867, pulpwood in 1886, and plywood in 1907.[16] Of these three, sawmills to produce boards, planks, and heavy squared timbers merit further note because they were the most consequential for Goryōrin in Naichi and, together with entrepreneurial pulp mills, were central to the exploitation of Hokkaido woodland.

For centuries, Japanese loggers had used broadaxes to square timber for rafting out of the forest, with sawyers in town using handsaws to cut the roughhewn sticks to consumption size and shape. The first mechanical saw was introduced in 1867, and a few water-powered sawmills were erected in following decades. The first sawmill for use in Goryōrin was a water-powered circular saw set up beside the Adera River in 1905, four years after construction began on the Adera's forest railway and just as construction of the Chūō Line was creating a rail link to the Nagoya market.

Four years later, in 1909, that mill was shut down as obsolete, displaced by a much larger one at Nojiri, on the east bank of the Kiso River just above the Adera's mouth. It used an eighty-horsepower steam engine to operate a band saw, a vertical saw, and a circular saw with rolling carriage. The mill also had a two-track cableway some 240 m in length that carried logs to the mill from the forest railway siding on the west bank of the Kiso. It swung them up over the river and across it to the mill's storage yard, which stood some twenty meters (about seven stories) higher than the siding. That move placed the logs, along with the mill's lumber piles, close by the roadbed of the advancing Chūō Line.

In 1911, after only a couple years of operation, the grand new mill was destroyed by fire. But in 1913, with the Chūō Line open and forest railway work starting along the Ogawa, it was rebuilt and equipped with an electric generator.[17] But shortly thereafter a sharp spike in the price of logs during World War I made it more profitable for the Property Office to ship logs directly to market by rail and sell them to entrepreneurial mills. Moreover, those mill operators were by then vigorously protesting the construction of government and Household sawmills. So the Nojiri mill and a smaller water-powered one a few miles south of there were shut down and sold. By then, however, the expanding rail network and growing number of entrepreneurial mills assured that any timber produced in Naichi Goryōrin could be transported and processed for sale elsewhere.

In Hokkaido, on the other hand, the Property Office never tried milling its own sticks. There, as noted earlier, lumber and pulp companies erected mills at strategic points and fed into them logs taken from Goryōrin, *kokuyūrin*, and entrepreneurial woodland, whether their own holdings or those of other forest owners. So the cessation of milling in Kiso marked the end of the Property Office's brief venture into sawmill operations.

Figs. 6-2a, 6-2b Sawmill at forest edge. These photographs, ca. 1980, show a small entrepreneurial sawmill sited at the base of timber-rich hillsides west of Tokyo. Nearby piles of logs for sawing and stacks of slabs (the outer edges of logs produced by milling) line the road along which trucks haul the logs from forest to mill.

LOGGING PROCEDURE AND PRODUCTION[18]

These changes in forest technology were intended to enhance logging output and Household income. Changes in the methods and organization of logging itself were also made, and for the same purposes. The cumulative effect of these changes, in terms of actual timber output and long-term potential, was in some ways stunning but in others ambiguous.

As noted in Chapter 5, the Property Office's original management plan of 1899 – which had the long-exploited forests of Naichi in mind and reflected both indigenous and current European silvicultural thought – envisaged a system of rotation blocks. Poorly-stocked sites would be cleared off and parcels hand-planted sequentially to establish cycles of even-aged monoculture stands that could then be thinned at proper intervals. A century or so later, the mature stand in each parcel would be removed in the final cut and a new generation of seedlings set out on the denuded parcel.

Later plans modified this system of thinning and final clearcutting, but from first to last, practice only partially conformed to plan. Most noticeably, logging, like afforestation, differed by region, and again the main regional distinction was that between Naichi and Hokkaido.

Regional Differences

In Naichi, clearcutting was the preponderant felling method. However, it slowly declined in importance, giving way to selective felling as work moved beyond the long-used hillsides into less accessible areas and as new silvicultural understanding was disseminated, especially from about 1920 onward. Thus, whereas eighty-five percent of logging in Naichi Goryōrin was recorded as clearcutting in 1918, by 1938 the figure had dropped to sixty-five percent (see Tables E-5 and E-6 for more figures).[19] Moreover, where clearcutting did continue to be practiced, mostly on poorly-stocked sites that needed thorough planting, the size of parcels was reduced and more shelter belts and seed trees were left standing.

In Hokkaido, by contrast, the Property Office's initial policy of clearcutting to produce merchantable logs failed from the very outset. For several years the Sapporo office pursued little active timber production because, with labor scarce and transportation poor, buyers were few and profits slender. Such logging of the island's old-growth forest as did occur was pursued initially as winter work by farmers, primarily on their leaseholds and mainly to clear land for agricultural use.

From about the time of the Russo-Japanese War, however, the earlier-noted establishment of sawmills and pulp mills expedited timber processing on the island. The work was handled by large lumber and pulp firms, notably Mitsui Bussan, Sapporo Mokuzai Kabushiki Kaisha, and the Fuji and Ōji pulp companies, which negotiated multi-year leases of designated tracts of Goryōrin (as well as *kokuyūrin*). Those arrangements led to a sharp increase in the logging of conifers for pulp and saw timber and broadleafs for charcoal and specialized products. And some clearcutting

Fig. 6-3. Plantation forestry technique. In this photograph of conifer *sanrin* west of Tokyo ca. 1980, one can see the thin lines of shelter-belt trees separating parcels that have at different times been hand-planted, nurtured (competing growth suppressed, restocked, limbed), thinned, and, as in foreground, finally clearcut for restocking.

was done, about 50,000 *chōbu* of Goryōrin by 1924. But, by and large, logging entrepreneurs in Hokkaido adhered to the *zenbatsu* or "gradualist" policy of "creaming" stands; i.e. taking the best and leaving the rest.

As years passed, foresters in the Sapporo headquarters acquired an ever stronger interest in developing operational plans and harvest policies that would assure them a sustained yield and proper stand regeneration. That orientation brought them into sharper conflict with the commercial loggers, prompting them to favor closer official supervision of logging operations. Gradually they negotiated more tightly regulated leases that permitted logging only within limits specified by contract and in line – supposedly – with the management plans that, by the 1920s, encompassed most of the island's Imperial Forest.

In specifics, then, the perceived shortcomings in logging practice differed by region. But in both Naichi and Hokkaido, Household foresters tried to improve performance by tightening and extending their control of the logging process. They did so, needless to say, because from first to last Goryōrin existed to fund the Imperial Household, and they saw those policies as the means of maximizing their estate's sustainable profitability. Tightening was a way to minimize the damage that others might do to the estate, and because logging was the single most destructive woodland operation, it was the one they needed to control most carefully. And extending their control to cover more steps in the lumber-processing operation seemed likely to increase the portion of total profit that would flow their way.

The Expansion of Household Logging

These considerations prompted officials to try to replace entrepreneurial arrangements with operations handled by their own supervisory personnel, technical experts, and locally-recruited labor crews. And they did make some progress in that effort. As Table E-10 indicates, Household-managed logging rose from less than 10% of Goryōrin totals in the 1890s to over 30% during the early 1930s. And the trend continued thereafter, rising to 38% in 1937 and 59% in 1947, just before Goryōrin merged with *kokuyūrin*.[20]

This trend to Household-managed logging occurred mainly in Naichi, but to some extent in Hokkaido as well. During the 1920s, officials at the Sapporo headquarters began organizing their own logging projects, initially to help meet emergency timber needs that followed the Kantō earthquake. They planned projects, hired, supervised, and paid crews, and arranged the disposal of their logs. They introduced the Caterpillar tractors mentioned earlier and during the 1930s pursued railroad and related construction, in the process gradually expanding their portion of the harvest.

Those efforts notwithstanding, the sale of stumpage still prevailed in Hokkaido. However, the method of sale changed. In earlier years, because competition for the business was weak, the Property Office followed customary practice that relied heavily on negotiated ("private" or "special") sales to locally established entrepreneurs or select companies such as Mitsui Bussan. By the twenties, however, publicly announced auctions and competitive bidding for contracts had become the norm. Whether that method was more profitable is unclear from the available figures (see Fig. 6-4 below and Table E-13).[21]

Trends in output

Regardless of whether the men who did the actual felling worked for entrepreneurs or directly for the Property Office, they still used crosscut saws, axes, pole hooks, and horses. And many were still nearby farmers doing off-season work. By the thirties, however, they were producing wood for more diversified purposes, including poplar (*doronoki*) for matchsticks, walnut (*kurumi*) for gunstocks, various broadleafs for plywood, and ever more varieties of wood for the pulp used in production of paper, fuel, and artificial fibers.[22]

The workmen, of course, were doing their best to piece together a living. From the Imperial Household's perspective, however, their labors were meant to provide it with a reliable income, and the Household's operatives tried to work out harvest schedules that would assure a reasonably stable and sustainable annual yield. In terms of total output they succeeded, with the annual cut rising steadily from about 450,000 m^3 during the 1890s to 1,900,000 during the 1930s (see Table E-10).

Whether this success could be sustained indefinitely, however, is less clear. Some measures of performance were less reassuring, although their significance is not entirely clear because of the earlier-noted expansion in acreage covered by management plans. To wit: the estimated volume of stumpage per hectare steadily declined, at least until 1938. Until the later 1920s, moreover, both the average volume of wood harvested per hectare and the annual increment of growth also fell (see Table E-7).[23] In part, these trends merely reflected improvements in record keeping; in part, the increase in acreage under active management, and in part, the establishment of more and more young plantation stands, mainly *sugi* and *hinoki* in Naichi. But in part they may well have signified overcutting.

If, indeed, loggers were consuming Goryōrin timber faster than the forests could replace it, their capacity to do so was dependent primarily on three facts. Most obviously, they had vast reaches of old growth timber to exploit in Hokkaido. Also, new modes of transportation made it possible to extract timber both there and in Naichi at unprecedented rates and from sites that had been nearly inaccessible just a few years earlier. And thirdly, a large, expanding, and diversifying market that sustained high timber prices until 1920 made logging a profitable enterprise for the Imperial Household. That market included both domestic and foreign consumers, and its growing diversity meant that more and more varieties of wood could profitably be extracted. This trend increased both the acreage worth harvesting and the completeness of the cut on specific parcels.[24]

On the other hand, some statistics may show that, in the longer term, Goryōrin policy was indeed attaining its goal of raising the level of sustainable output. To wit: Table E-7 reveals that during the 1920s and into the thirties, the decline in forest productivity slowed, with some trends even reversing in areas of Naichi. To some extent, of course, this apparent improvement simply reflected the fact that less new acreage was being added to the area under management. And to some extent it reflected the curtailment of Property Office logging in response to the 1920s declines in lumber prices that resulted from economic recession and import competition, mainly from North America and Karafuto.[25] Nevertheless, the figures could still mean that on balance the stumpage losses to logging were beginning to be more than offset by the gains due to years of natural regeneration and afforestation effort.

Stumpage figures for 1939 and 1946 also seem to suggest the longer-term effectiveness of Property Office policy. Although wartime and postwar demand stripped Japan's woodland of everything of size that loggers could reach in the time available, Goryōrin stumpage figures for those two years seem to tell a story of growing biomass. They report some 550 million *koku* of stumpage in 1939 swelling to 645 million by 1946 (see Table E-12). These figures suggest that the growth rate in those plantation and natural stands too young to be merchantable – *hinoki* and other needleleafs most notably – was more than sufficient to replace the stumpage volume being lost to rampant felling.

It may well be, however, that the figures are deceptive. With ample manpower available in 1946, Household foresters may simply have been able to obtain a more complete and accurate count than in 1939. Or the 1939 totals may have been understated to reduce government pressure on Household foresters to harvest, and those of 1946 overstated to boost the apparent asset value of Goryōrin prior to its re-inclusion in government woodland.[26]

In short, from the 1890s to around 1940, Goryōrin logging output rose about sixfold (see Tables E-10 and E-11), and the preponderance of information suggests that it served the purposes Meiji leaders had intended. Whether or not the Imperial Household was, in the longer term, living beyond its silvicultural means seems less clear.

THE PAYOFF

Ultimately, of course, the purpose of Goryōrin was to produce timber whose sale would help finance the Imperial Household. During its half century of existence the timber output of Goryōrin rose steadily. And marketing agents had even more success in converting the timber into income. Specifically, as Fig. 6-4 and Table E-13 attest, the annual net profit from Goryōrin exploitation soared from around 234,000 yen in the later 1890s to some 6,000,000 yen during the 1920–1930s. A portion of that disproportionate monetary gain was illusory, of course, a reflection of general inflation. But much of it was real, flowing from improvements in production efficiency, expansion in total output, and the exceptional price rises experienced by timber.

To look at that financial record a bit more closely, during its first couple of years Goryōrin operations yielded annual losses. But from 1892 onward it turned a profit, becoming for several years a quite steadily growing source of income.[27] Beginning around 1900, however, the income flow from the expanding cut grew increasingly erratic as Japan's economy became more deeply enmeshed in the global politico-economic disorder of the day. That disorder – notably gyrations during the Russo-Japanese War, a wild boom during World War I, a collapse in 1920, an erratic recession during the twenties, the trough of the Great Depression in 1929–33, and the war-related rebound of the late thirties and early forties – produced radical swings in the market price of lumber.

The resulting fluctuations in gross income were not significantly dampened by swings in production costs, with the result that Property Office profit margins jumped and plummeted wildly, reminding one of the boom and bust cycles that plagued Tokugawa-era lumbermen. Income from the sale of unwanted parcels of *goryōchi* further exacerbated the fluctuations, or occasionally smoothed them. As a consequence of these factors the history of a reasonably steady growth in forest output is substantially obscured by the surface picture of a chaotic income stream.

When compared to *kokuyūrin*, moreover, Goryōrin was highly productive and profitable. As revealed by the Grand Totals in Table E-1 of

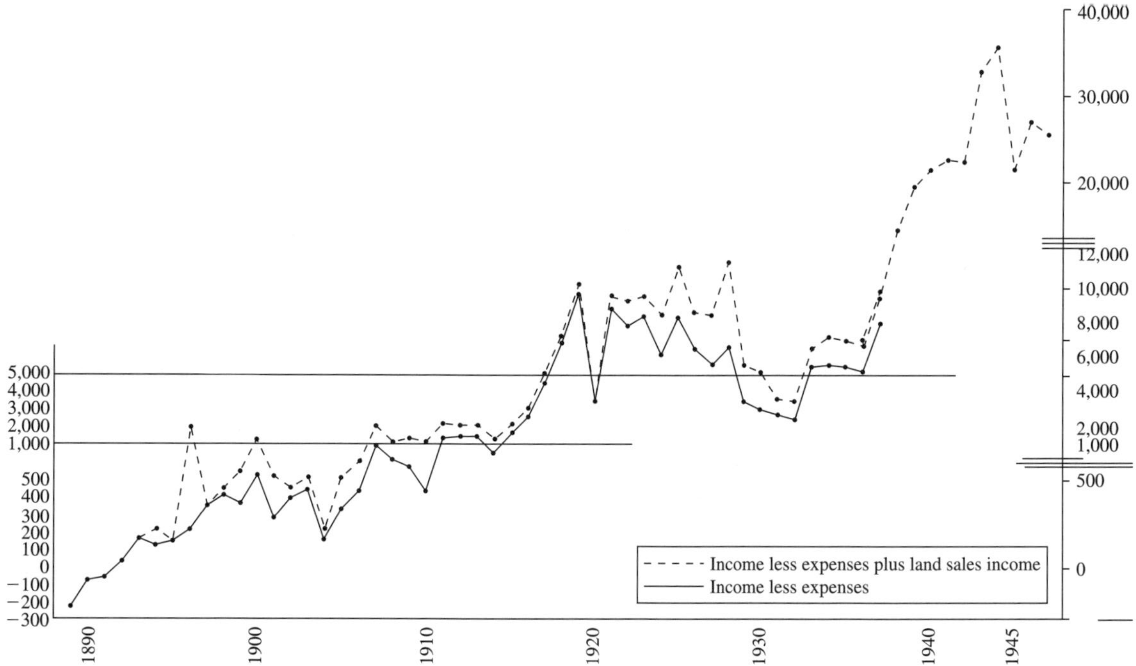

(Source: Based on figures in Kuroda Hisata, *Tennōke no zaisan*, pp. 248–9, Table 2-1 (for years 1889–1937); Hagino Toshio, *Nihon gendai rinsei no sengō katei*, p. 60, Table 6 (for years 1936–47).)

Fig. 6-4. Trajectory of Goryōrin Profits, 1889-1947 (in 1000 yen)

Appendix E, in 1939 Imperial Forest contained only about 18% as much acreage and stumpage as *kokuyūrin*. As Table E-14 reveals, however, during the decade 1936–45 it generated nearly a third as much yearly gross income (52.6 million yen). Furthermore, thanks to greater efficiency of output, an even higher proportion of its gross converted into net profit (40% as against 23% for *kokuyūrin*). As a consequence, although Goryōrin had only 15.1% of the acreage of *kokuyūrin* and Goryōrin combined, it yielded 36.6% of that woodland's combined net profit.[28] The initial care with which parcels had been selected for inclusion and the subsequent decades of silvicultural improvement seem to have paid off.

When looked at in terms of the Imperial Household's total budget, moreover, Goryōrin proved to be a major and growing element (see Appendix B). Whereas it had lost money before 1892, it provided 19% of the Household's total income in 1916 and 31% by 1940. Cumulative investment in Goryōrin was so massive, moreover, that by 1946 its assets had grown to constitute some 70% of the Household's total material wealth.

This long-term growth in the centrality and fiscal value of Goryōrin to the Imperial Household notwithstanding, after Japan's defeat in 1945 the basic rationale for its existence disappeared. The Constitution issued in November 1946 declared in Article 88 that,

> All property of the Imperial Household shall belong to the State. All expenses of the Imperial Household shall be appropriated by the Diet in the budget.[29]

Both the logic and structure of Goryōrin had been repudiated. It could not thenceforth be distinguished from *kokuyūrin*. So, in March 1947, new legislation combined the administrations of *kokuyū* and *goryō* woodland, ending the separation that had created so much difficulty for so many people only a half-century earlier.

7

Goryōrin: An Assessment

By the spring of 1947, Goryōrin had come and gone, dissolved, as it had been created, for reasons of political ideology. And within a few decades it was largely forgotten. But in its own day it was a significant part of both Japan's forestal and royal history, and today it still sheds light on those experiences. It also gives us further insight into ways in which the British and Japanese historical experiences have and have not been similar.

REFLECTIONS ON GORYŌRIN

Aspects of Goryōrin history – its founding, difficulties, and overall performance – seem to merit some last reflections, as may a brief "might have been," had Imperial Forest not been summarily abolished in 1947.

Its Founding

At the time of Goryōrin's creation around 1890, the decision by Meiji leaders to establish an independent fiscal foundation for the Imperial Household was not unreasonable. As inheritors of the 1868 victory, after all, they were the leaders of a conquest regime and had ample reason to guard the heart of their claim to legitimacy against both rivals for power and the discontented among those whom they ruled. Moreover, much political rhetoric of the day – whether employing ideas and information from Europe or from their own political heritage – helped validate the whole notion of a financially robust imperial institution.

More specifically, considering both the precedents of Japan's own hereditary elite and current practice among the crowned heads of Europe, it was reasonable to give the royal house a substantial landed base. Furthermore, given the expansiveness of Japan's forested terrain and the intensive use to which the remaining land was already being put, once the fiscal potential of timberland was understood, it made good sense to center that landed estate in woodland. Finally, such an arrangement fitted nicely the new legal principles that Meiji leaders were appropriating from Europe, in particular the notions of exclusive property ownership and property taxes. These concepts provided the government with a doctrinal basis for repudiating the whole complicated and ambiguity-ridden legacy of use rights and obligations that blanketed Naichi, replacing it with a seemingly clearcut system of property rights that would permit the creation in perpetuity of an unencumbered imperial estate.

Its Difficulties

Precedent, logic, and legal niceties notwithstanding, from the very outset Goryōrin encountered difficulties. The mere facts of change itself, of new demands on old woodland, of demands by uninvited outsiders, and of too much demand on too few resources were ample grounds for conflict and difficulty. Much Goryōrin in Naichi was neither as productive nor as extensive as its creators seem to have presumed, and much was already being used more fully than they evidently realized. Then, when the Property Office addressed those problems by agreeing to one or another type of rental arrangement, the Imperial Household lost much of the fiscal value of its estate.

Regarding that last point, those rental arrangements assured many villagers of access to basic necessities – fertilizer, fodder, fuel, and myriad lesser items. And village dependence on this type of forest yield did not decrease during Goryōrin's lifetime. Quite to the contrary, Japan's unprecedented population growth (from about 32 million in 1870 to more than 70 million in 1940) expanded the total demand for such customary forest products, and villagers fought hard to get enough to survive. Even when they were able to harvest the goods in ways that did no environmental damage (and given the rising level of demand, too often they couldn't), village income levels generally were so low that the Imperial Household could not squeeze a significant rent out of the harvest. And land devoted to production of these goods could not be utilized in other ways more profitable to the Household.

Moreover, villagers (larger landowners and local lumbermen in particular) discovered new ways to turn more and more woodland to advantage as new uses for forest produce created new demand and thus new market opportunities. These new types of demand included mulberry leaves for producing export silk; charcoal and fuel wood for powering steam engines and smelting metals; pulpwood for paper and chemicals; construction wood for railroad ties, mine timbers, vehicles, vessels, buildings, bridges, and so on; and specialty wood for a grand variety of purposes, including tea chests, matches, export bric-a-brac, gun stocks, warships, and, by 1940, airplanes. So villagers had both a need for more goods in their own lives and the opportunity to benefit from the marketplace if they could harvest stands or exploit lands at will, including those the Imperial Household claimed as its own. And few among those villagers wished to share their gains with the hereditary elite.

On Naichi, then, much of the land that became *goryōchi* was already being intensively exploited, and its local exploiters justified their claims primarily through the sturdy, long-sanctioned principle of communal use-rights. The imported notions of exclusive property ownership and property taxes that the government was promoting were, for all their seeming up-to-dateness, singularly at odds with the existing forest situation. So their application only complicated the Household Ministry's other woodland difficulties.

In the end, it is true, the Ministry did gradually impose its control, organize its realm, and make real gains in timber output and the fiscal value of its holdings. But it did so only by a long, painful process that involved accommodation and compromise as well as forceful insistence; trading land for cash; gradually and reluctantly abandoning the most troublesome parcels in return for a freer hand on what remained; and finally, in 1920, offering villages payments-in-lieu-of-taxes. Primarily because of that tortuous process, by the 1930s, the Imperial Household ended up with a woodland estate that was far smaller and far less elegant than the Fuji-centered realm originally envisioned by its creators.

Its Performance

Mindful of Goryōrin's difficulties, one might plausibly argue, on the blissful basis of twenty-twenty hindsight, that from first to last it was a stupid idea. Not only did it complicate the task of national woodland management, but it also pitted the Household against a large population of villagers. In addition, it grounded the imperial fisc in one of Japan's most erratic marketplaces, that of the lumber industry, with its long history of feast or famine. More basically, one could assert that it was anachronistic to try to anchor imperial well-being in an agricultural economy – the cropping of woodland – even as Japan was marching quickstep into an industrial economy.[1] So how well did it perform, whether as a fiscal system or a silvicultural one?

1. *Fiscal Performance*

In actual fact, of course, from the 1880s to the 1940s, the major element in Household assets was not woodland at all but, as Appendix B indicates, industrial-age stocks and bonds, which generated about half of its total income. However, the stocks were susceptible to the radical ups and downs in valuation that affected all such market investments during those decades. So they provided a fiscal base that was even less reassuring than Goryōrin, all its difficulties and wild swings in profitability notwithstanding. Worst of all, much of the Household's investment portfolio was in empire-related banks and other enterprises in Taiwan, Korea, and Manchuria, and, in the end, military defeat destroyed their value, leaving Goryōrin as the Household's main surviving asset.[2] Woodland was not, it must have seemed in 1944–46, such a stupid idea after all.

Considering, also, how much profit the Household reaped even from its much-reduced Goryōrin, one suspects that it never really needed the vast acreage it originally received. Rather, it seems likely that its creators underestimated how much the market price of timber would rise relative to other prices and hence how much value was embedded in every *chōbu* of woodland they controlled.

For all the problems it presented, then, Goryōrin proved to be a relatively advantageous fiscal foundation. Indeed, it generated ever-higher levels of profit, even allowing for inflation, which meant that it was, by definition, a success. As noted in Chapter 6, however, that success mainly reflected high demand for wood, the near absence of foreign competition before 1920, and one-time harvesting of old growth. How sustainable it would have been in the longer term is hard to say.

2. *Silvicultural Performance*

In terms of silvicultural theory and practice, the record on Goryōrin seems reasonably creditable by either domestic or global comparison. In Hokkaido the prevailing old-growth stands of mixed timber were gradually surveyed and, in principle, subjected to management plans that mostly called for selective felling and natural regeneration. In practice, however, most of the island's Goryōrin was logged off in a gradualist manner that in essence creamed the forest, while planting practice was sufficiently faulty that most of the woodland was in effect left to regenerate as best it could. The basically exploitative character of this practice notwithstanding, however, it appears that woodland on the island was extensive enough and the rate of felling in this pre-chain-saw era slow enough so that its Goryōrin as a whole remained biologically vigorous, if messy, at least until well into the Pacific War.

In Naichi, where Goryōrin contained many long-worked plantation stands and brushy hillsides, theory initially favored clearcutting and hand-planting of diverse species, a policy that made sense on such sites. However, hand-planting was costly, which encouraged reliance on natural restocking. But the latter's results often were disappointing, so policy preference shifted back and forth. By the 1920s, planners were encouraging a mix of hand-planting and natural regeneration, promoting it strongly in conjunction with the maintenance of multi-aged, mixed forest.

Practice, of course, was imperfect, with planters in Naichi setting out more and more *hinoki* and fewer and fewer other species, in the process producing ever more even-aged monoculture stands on the more accessible slopes. And, despite increasingly-sophisticated techniques for estimating yields and planning forest management, felling proved erratic. The ideal of a stable, sustained yield suffered due to sieges of emergency harvesting, notably those related to the Russo-Japanese War, the Kantō Earthquake, and the Pacific War, and, at the other extreme, a reduced level of logging during the 1920s, in the face of a market decline brought on by domestic recession and unprecedented timber imports.

Given the ecological costs and risks of even-aged monoculture stands, particularly in terms of vulnerability to disease and loss of woodland biodiversity, one might question (in Goryōrin as in the world's other industrial-age forest systems) the wisdom of establishing such stands, notably the plantations of *sugi* and *hinoki* found in Naichi. Of course, Goryōrin occupied only about six percent of Japan's woodland area and consisted of parcels scattered widely about the realm, so even if fully stocked with such trees, it presented by itself no particular ecological threat. However, its parcels were found among the much more extensive holdings of *kokuyūrin*, and on those holdings – as well as many parcels of *min'yūrin* – planters also favored *sugi* and *hinoki*. In consequence, the realm was trending more and more toward large, contiguous acreages of even-aged monoculture stands. The trend was, in essence, another of the myriad global trends whereby industrial society is undermining the very biosystem that sustains it.

Still and all, as of 1940, Goryōrin, along with Japan's other woodland, continued to be characterized by widespread stand diversity. And planting theory of the day seemed to favor its preservation. However, considering the strong trend to *hinoki* culture in actual afforestation projects, the main factor behind the persistence of this stand diversity may have been the widespread failure of those planting efforts rather than the presence of foresighted planting theory. In any case, the problem of excessive even-aged monoculture forest mostly came later, following the massive postwar restoration of Japan's woodland, which produced the vast sweeps of park-like conifer forest of the late-twentieth century.

And if Goryōrin Existed Today?

If one opts for the happier assessment, concluding that Imperial Forest was not only profitable, but also productive, biologically healthy, and not unduly oppressive of society, then one might argue that it simply had the misfortune of being killed before it could fully prove itself.[3] After all, the tree planting that increased steadily after the 1890s was creating stands which only began reaching harvest age during the decade or so before Goryōrin's dissolution. Their growing value, as we noted earlier, was suggested by increases in annual increment after 1928 (see Table E-7) and in stumpage after 1939 (Table E-12). And as a portion of *kokuyūrin* during the 1950s and sixties, these maturing stands yielded much lumber for postwar reconstruction and economic expansion.[4]

Indeed, in a number of ways the decades since Goryōrin's dissolution have been far more favorable to the venture than were earlier ones. Most importantly, from the 1950s onward village demand on woodland dropped precipitously for a number of reasons. Specifically, a society-wide shift to reliance on fossil fuels drastically reduced the use of firewood and charcoal. Moreover, commercial fertilizers replaced mulch, while motorized vehicles and machinery obviated pack and work horses and the pasturage they required. In addition livestock-feed imports replaced fodder for meat and milch animals, and synthetic fibers nearly destroyed the silk industry that had required mulberry trees. Finally, the use of aluminum, plastics, and other materials eliminated most demand for thinnings and small timber of the sort that villagers once produced.

Less obviously, demands on woodland have been eased since the 1950s by a radical expansion in industrial-age forms of employment. Together with burgeoning food imports, this expansion has more than offset the effects of continuing population explosion (from some 70 million in 1940 to 127 million in 2005). Cumulatively, these two trends have reduced sharply the role of agriculture and rural production in the Japanese economy, causing mountain villages to shrivel all across the realm and greatly easing village pressure on all types of forest.

Because of these developments, from one end of Japan to the other villagers nearly stopped using hillsides and marginal plots of pasturage or arable. Instead, they planted many plots to timber seedlings and let others revert to wild growth. As a corollary, after the 1950s, the once-burning issues of *iriaichi* and woodland use rights gradually disappeared from view. Many woodland areas that once had been contested parcels of Goryōrin thus became available for growing timber, completely free of competing uses. This outcome suggests that the only problem with Goryōrin was that leaders begat it and then buried it a few decades too soon.

Other factors, however, suggest a very different conclusion. Specifically, as Japan's industrial economy boomed after 1960 or so, labor became scarce, especially for manual work in hinterland areas, and wage rates rose sharply. Because of Japan's topography, logging remained labor intensive, even with chain saws and other modern gadgetry, so rising wages drove up its cost. Meanwhile, cheap fuel oil had become available and huge ocean-going freighters proliferated. So maritime shipping costs dropped and foreign trade mushroomed. Also, after about 1970, a chronically favorable balance of trade pushed up the exchange rate of the yen from a fixed 0.28 U.S. cents per yen to a fluctuating 0.7–1.10 in the years around 2000.

Together with the rise in Japanese labor costs, that currency shift and the explosion in maritime trade transformed the country's timber market. In most product areas, it became cheaper from the seventies onward for Japanese wood processors and consumers to use imported timber than to purchase domestically-grown stock. That situation persisted thereafter and shows no sign of changing in the foreseeable future. In consequence, the timber-growing forests of Japan are no longer an appreciable source of profit for those who control them. Of course, it seems likely that global woodlands will become so depleted at some date in the not-too-distant future that timber prices will escalate enough to make Japan's forests again a source of profitable production. Then the Imperial Household may well regret its loss of timberland.

Today, however, if Goryōrin still existed, it would be an albatross about the imperial neck, consuming the imperial fisc rather than supporting it. Inadvertently those who stripped the Imperial Household of its autonomous fiscal foundation in 1946 saved it from a host of later difficulties. As seen from the perspective of the year 2005, then, it is one of history's nice little ironies that the reformers of 1946, in their zeal to make Japan into an Anglo-American-style democracy, saved the imperial institution from later grief. Rather, they gave it a fiscal foundation that may prove far more durable than the one devised by Meiji reformers in the 1880s.

BRITAIN AND JAPAN AGAIN

Finally, to return to the comparative agenda adumbrated in the Preface, a number of points stand out when one reflects on Goryōrin in terms of Britain's Crown Lands. Again one notes both similarities and dissimilarities.

Similarities

Most obviously, but perhaps least consequentially, much as Crown Lands became an element in the government fisc during the age of George III, so Goryōrin in 1947 returned, after its short, awkward life, to its initial role as a part of the Japanese government's woodland and fisc.

Other similarities are of much more fundamental import. In both societies, long-term population growth gradually intensified the pressure on woodland of all sorts, and during the past two centuries that intensification has been reinforced by the development of new technologies – notably railroads, sawmills, and the internal combustion engine. Under their impact, but sooner in Britain than Japan, more and more woodland became accessible to exploitation and fell victim to the saw, subsequently evolving into mixed secondary growth, plantation stands, or non-forest as circumstance dictated.

As woodland shrank and deteriorated, forms of forest regulation and regenerative forestry gradually appeared in both Britain and Japan. In both, those practices had domestic origins long antedating the nineteenth century, but also in both, German silviculture influenced policy from about the 1870s onward. And as it did, biological considerations of ecosystem maintenance gradually displaced legal questions of property ownership in the determination of forest policy.

Dissimilarities

Such similarities notwithstanding, however, the differences in these royal forest experiences still appear more compelling. Differences in the impact of topography seem most important, but dissimilarities in the historical impact of overseas forests and domestic "enclosure movements" also merit attention.

1. *Topography*

The key consideration, of course, is Japan's extremely mountainous character. It has meant that no matter how badly Japan's rulers and people wished to convert woodland to arable, they could do so only to a severely-limited extent. In consequence, a great deal of woodland survived – and still survives – in Japan. And, before 1947, it was able to play a valued role in the royal fisc, a role woodland had long since ceased to play in Britain. As we have seen, during Goryōrin's short life, the Imperial

Household's interest in forest output as a source of income never flagged, and it served the Household well indeed.

However, that very mountainousness has meant that, in Japan, erosion and flooding have been, from the outset, major consequences of careless deforestation. Much more than in Britain, therefore, protection forestry became central to Japanese woodland management. After 1879, concern about such flooding was a persistent factor in determining which woodland to "privatize" or hold under government control. And that consideration also played a key role thereafter in determining silvicultural policy on both Goryōrin and *kokuyūrin*. Moreover, the intensifying attempts by government during those decades to regulate the use of all woodland, whatever its legal status (which we noted en passant), sprang more from the wish to minimize environmental damage than from a desire to maximize forest production.

In Britain, on the other hand, the much gentler terrain meant that, as centuries passed, a far larger portion of the realm could be, and was, converted to agriculture. So forest long ago lost the capacity to be a major element in the royal fisc. Moreover, the central problem created by expanding deforestation and conversion to pasturage was scarcity of woodland yield, not erosion and flooding, and as a corollary the central goal of forest rehabilitation was to increase timber output, not to minimize downstream damage.

2. *History*

The forestal ramifications of these topographical dissimilarities then helped shape the way a couple of noteworthy historical developments worked out in the two countries: the rise of wood imports and the process commonly called an "enclosure movement."

Regarding imports of wood, as noted in the Preface, foreign woodland played a role in Britain centuries before it did in Japan. With the rise of empire in Elizabethan times, the monarchy acquired a need for timber with which to build ships of war and commerce. And a major function of Crown woodland was to provide the walls, decks, masts, spars, and knees of countless ships in His or Her Majesty's Navy. However, as early as the 1650s, the Crown found it lacked the trees it needed for tall masts. And a more general shortage of oak emerged soon afterward.[5] All the forests and hedgerows of Britain, whoever owned them, could not meet the growing demand, so leaders looked elsewhere.

They found the timber they needed overseas, and eventually those imports became sufficient to meet the requirements of both marine construction and domestic life. As that happened, it reduced pressure on the monarchy to restock forests, making it easier for farmers to convert yet more acreage to pasture and arable.

In Japan, by contrast, a policy of firmly limiting foreign trade, which was sustained from around 1600 to 1850, minimized the demand for large ships and for the timber they required. But the policy also prevented nearly all timber imports, which meant that practically all domestic wood consumption had to be satisfied from domestic forest growth. And it was. Essential needs were met initially by the felling of old-growth stands and, after 1700 or so, by a complicated combination of rationing and regenerative forestry. Timber imports did not appear until the 1920s.

When those imports finally did appear, moreover, their impact on Japan's woodland was very dissimilar to that of Britain. Whereas the imports to Britain had freed up woodland for conversion to other uses, in Japan they served mainly to undermine the economic value of hillsides that had little alternative human use. Far from making land available for tillage, or otherwise offering a boon to villagers, the imports deprived many of a lucrative sideline. The environmentally alert will see much to applaud in this situation; lumbermen, woodland owners, and forest workers did not see the positive side during the 1920s – and do not today.

Regarding "enclosure," in both societies enclosure movements effected profound changes in land-control arrangements. Both movements sought to end old communal use rights, and in due course they did so. In Britain, that movement occurred from around the fourteenth century onward, largely in conjunction with the spread of sheep grazing.[6] In Japan, enclosure was a much more recent phenomenon, occurring mainly from the 1870s onward, the so-called *kan/min* division.

Both movements did change the systems of land control, but they led to very different outcomes in terms of land usage. In Britain the enclosed areas mostly became pastureland on large private estates. In Japan the enclosed areas mostly became woodland, whether Goryōrin, *kokuyūrin,* or entrepreneurial timberland. In another of history's little ironies, however, due to other factors those trends neither intensified the British need for woodland nor eased that need in Japan.

To explain, in Britain the conversion of woodland to pasturage increased greatly the numbers of animals whose manure could fertilize arable, reducing villagers' interest in forest mulch. Meanwhile, the transition from wood fuel to peat to ground coal sharply reduced human dependence on woodland as a source of fuel. Concurrently, the rise of urban industrial employment and the exodus of people to settle abroad reduced population pressure in the hinterland. And finally, the resort to wood imports and, after the 1860s, the replacement of wooden ships with iron eliminated the monarchy's need for forests. Together, these trends meant that by the end of the nineteenth century, even though most woodland had disappeared, neither rulers nor ruled had compelling reason to lament the loss. Aesthetic or ethnic sensibilities might be wounded, but the well-being of few households was at stake.

In Japan, on the other hand, the archipelago's mountainous character, with its great capacity for erosion and downstream damage, gave both government and the Imperial Household powerful reason to restrict public use of the woodland they had acquired through enclosure and to nurture it as both protection and production forest. However, woodland still remained important to villagers because, as noted above, the principal changes that would finally reduce their dependence on it – the replacement of wood with fossil fuel, of mulch with chemical fertilizers, of mulberry-based silk with artificial fibers, of fodder with imported feed grains, and of lumber with other construction materials – occurred from the 1950s onward, after the dissolution of Goryōrin. In consequence, the existence and arrangements of Goryōrin (and *kokuyūrin*) presented major difficulties to many households and remained a major source of friction between Japanese monarchy and people at least until those changes occurred.

In the end, then, perhaps the most charming irony in our story is that in Britain, even though woodland had largely disappeared by the twentieth century, what remained was of such little material consequence to British life that forest issues were not a source of severe and sustained conflict between rulers and ruled. In Japan, however, even though the realm was far more richly forest-clad, competition for the yield remained so intense as to constitute a central focus of ruler-ruled conflict until 1950 or so, as exemplified by the vicissitudes of Goryōrin.

PART II

8

The *Kan/Min* Division of Woodland in Early Meiji, 1871–76

To put the matter boldly, the Meiji government's 1870s–1880s program of woodland reorganization created the greatest conflict between government and people in the entire history of Japan. It was in essence a contest for control of the realm. As adumbrated in Chapter 1, it pitted the government (*kan*) against the 80 percent of Japan's people (*min*) then living in the villages of Japan Proper (Naichi), nearly all of whom had a direct stake in the issue.[1] Moreover, each side was convinced that it had the more righteous and more important claim. Hence the stubbornness of the struggle.

The contest began around 1871, and it became most heated during the 1880s. There was no real ending date; the struggle just gradually petered out between the 1890s and 1920s or thirties. One can, however, point to 1889 as the year when the government in Tokyo essentially completed its effort at forest consolidation. And the basic character of the *kan/min* struggle was established by 1876. So here we focus on those formative years of the seventies.

In the outcome each side secured a piece of the pie. But the path to that outcome was extraordinarily convoluted and confusing, and it was extremely difficult for everyone involved. The woodland, too, paid a high price, with decades of effort required to undo the damage.

DIMENSIONS OF THE PROBLEM

The issue of woodland organization was in part a by-product of early Meiji land-tax reform, an undertaking that encompassed both arable and woodland (as well as urban property). But in greater part it sprang from changing perceptions of government need and woodland value and the changing policies that resulted.

Both the land-tax and forest issues have been extensively studied by Japanese scholars.[2] And solid essays on the land-tax reform have appeared in English.[3] These latter, however, have focused on the reform of arable land, which constituted about 16 percent of Japan's land surface, with interest primarily in the 9 percent that grew wet rice.[4] They have not addressed the more thorny, conflict-ridden, and long-drawn-out part of the reform effort, which related to the non-arable 70-odd percent.[5]

But why, one may reasonably ask, was woodland (*sanrin gen'ya*) such a bone of contention? After all, most of it was found on hillsides so steep that even a goat would find them daunting.[6]

In part the answer lies in the very fact of that steepness. Should anyone mess carelessly with those hillsides, they might produce devastating erosion, landslides, and downstream flooding that could – and all too often did – ruin fields, homesteads, and villages. Contestants disagreed, sometimes very strongly, on what sorts and what extent of encroachment on hillsides were acceptable and on who – in particular whether locals or outsiders – would do the most satisfactory job of regulating it.

Even more central to the conflict than this issue of "protection forestry" were issues of "production forestry." These constituted a perversely interconnected set of problems. (1) There existed a basic disjunction in the woodland-use priorities of people and government, one that sustained a fundamental disagreement on what type of woodland was desirable and, hence, on how it should be managed and, therefore, by whom. And that disjunction was exacerbated after 1868. (2) There emerged a basic juridical issue of who had what rights to use which woodland, how those rights would be defined and determined, and whether such woodland use should be taxable or tax-free. (3) There was, finally, a cluster of issues relating to the character and quality of officialdom's handling of woodland matters during these years. Let's examine these three topics seriatim.

On Priorities of Use

Woodland can accommodate a wide array of human wants, and it did so in Japan. Within the diversity of uses people had for woodland, a clear division existed between the priorities of rulers and villagers. The latter needed fodder for animals, an array of foodstuffs for themselves, water and green fertilizer for fields, brush wood for fuel, and mostly small-size timber for construction. So they needed *gen'ya,* meaning terrain that sported substantial understory along with some larger trees: mixed growth rich in bamboos, other grasses, forbs, and vines, as well as deciduous broadleafs (such as oak and beech) that produced ample coppice growth.

Rulers, on the other hand, needed large-size construction timber to maintain their cities, towns, and monuments to power and privilege (castles, palaces, mansions, temples, and shrines). And they needed high-grade fuel-wood stands to produce the charcoal and faggots used in both daily life and diverse industries. So they wanted *sanrin,* which meant high forest rich with timber-producing trees, notably such conifers as cryptomeria and Japanese cypress (*sugi* and *hinoki*). And those treasured conifers were masters at shading out and destroying the very understory that villagers required.

Fig. 8-1 *Sanrin.* A stand of young *sugi,* well thinned and trimmed, with low sunlight flooding in from adjacent clearcut area off right. In mountains west of Tokyo, near Ōme. (From author's collection.)

Fig. 8-2 *Gen'ya.* Oak coppice (sprouted from stumps) with understory of *sasa,* a small bamboo, visible on far side of wood road. *Sasa* in foreground has been cut and removed for mulch. On terrace deposit just across Tama river south of Tokyo. (From author's collection.)

This disjunction in wants, and hence in preferred woodland type – *sanrin* vs. *gen'ya* – raised the issue of who should control the land so as to assure its "proper" management. Whereas there was long-standing agreement that arable land was best left in the hands of villagers because rulers and ruled were in essential accord about what should be done with it, the disjunction in preferred uses of woodland led directly to disagreements about who should control it – *kan* or *min*. And during the seventies, that dispute intensified, becoming central to the struggle over woodland.

One factor exacerbating the issue of woodland priorities during the seventies was the discovery by Japan's leaders of a host of new construction wants – notably warships and wharves, railroad ties and rolling stock, mine timbers, telegraph poles, factories, and other buildings – that sharply increased the total demand for timber and commercial fuel wood. That increase in demand prompted government to want more and better *sanrin*, but many of the sites that could best grow large-scale timber were the lower, more accessible hillsides that villagers valued most highly as *gen'ya*.

It should also be noted that as the decade advanced, more and more villagers and entrepreneurs became aware that new demands were creating a profitable seller's market for both timber and fuel wood.[7] So one factor in both the accelerated scramble for control of exploitable woodland and the reckless felling of these years was a desire to exploit that market, whether out of desperation, greed, or grand design.

On "Use Rights" and "Ownership"

The issue of control, needless to say, was essentially a question of power. But in Japan, as in other complex societies, questions of power are commonly addressed through structures and principles of law, and this case was no exception.

To elaborate, for centuries a wondrously complex system of shared-use rights (commonly called *iriaiken*) had accommodated the basic facts of woodland's diverse capabilities, people's many needs, and the disjunction between village and government priorities.[8] The system specified (with varying degrees of preciseness) who might do what, where, when, how, with what tools, and how much yield, thereby assuring a multiplicity of identified users the right to extract designated goods from particular sites. It was a shared-use (or multiple-use) system that fitted the existing ecology very nicely.[9]

Moreover, because forest biota habitually display a brazen indifference to the niceties of human boundary-marking, these shared-use arrangements came to be applied with little regard for official categories of land. Whether the acreage in question was formally designated household land, communal land, or government land ("lord's forest," most commonly called *ohayashi*), which were the three major official categories of Edo-period woodland, *iriaiken* could be found on all of them.[10]

During the 1870s, Meiji leaders undertook – for purposes of tax enhancement initially and firmer forest control later – to replace these shared-use arrangements with a uniform, countrywide system of exclusive property "ownership" (*shoyūken*). To foreshadow – and oversimplify – a story marked by change, compromise, confusion, and inconsistencies, the new system would differentiate between "government land" (*kan'yūchi*), which was to be both tax-exempt and free of all shared-use obligations, and "people's land" (*min'yūchi*), which was to be taxable and subject to the owner's autonomous control.[11] In adopting these goals, the authorities created a situation in which a vast number of important interests were threatened while little more than the interest of the reformers seemed likely to benefit. And the anticipated losses of fodder, foodstuffs, fertilizer, fuel, and timber were not trivial. Faced with the prospect of losing basic necessities, villagers took such measues as they could to prevent that outcome.

By nullifying use rights, "ownership" became the enemy, and from a villager's perspective that enemy had two faces. One was that of the rich merchant or landlord who, if he acquired woodland as his own, could then do with it as he chose, while excluding others, denying their customary use rights, or charging them for any use of it. The other, greater enemy was the government, which could – and did – claim for its own exclusive use vast acreages that villagers had hitherto utilized for their diverse purposes. So, regardless of whether "*kan/min* division" (*kanmin'yū kubetsu*) – the process of assigning all woodland to the basic categories of tax-free *kan'yūchi* or taxable *min'yūchi* – yielded people's land or government land, most villagers saw the policy as a threat to their interests.

Even those villagers who became owners of small woodland parcels had little reason to welcome that outcome. After all, during the Edo period there had been no tax on woodland per se, and the miscellaneous taxes (commonly called *komononari*) on its various market products constituted an occasional and negligible burden. The new woodland property tax, in contrast, although extremely modest, meant that an owner faced an annually recurring and inescapable charge for use of land from which he hitherto had extracted home-use materials gratis.[12] And few of the newly created, small-scale owners seem to have rejoiced at the "capitalist" thought that they could now charge their neighbors for harvesting yield that hitherto had been theirs for the taking as per shared-use arrangements. Nor did they find pleasure in knowing that the land was now theirs to sell since doing so would merely prove that they had failed to preserve their patrimony and pass it on to the next generation.

On Officialdom's Performance

Thirdly, characteristics of official procedure added to the tension. One problem was the limited understanding that officials brought to their task. Most of them were urban samurai with little grasp of either village life or the existing woodland system and how it worked in all its complexity and regional variety.[13] Nor did they have more than a vague sense of how their intended new system of woodland organization could or ought to be implemented or of the effects that the planned changes would have on the lives of either rural folk or the woodland itself. So they learned as they legislated, by trial and error. And in the process they confused, frustrated, frightened, and angered people all across the realm.

In addition, Tokyo's forest policy was in flux throughout the decade. It became entangled in irreconcilably conflicting purposes, most notably whether to sell government land for income or hold it – indeed, expand it – to grow timber. It also changed as the level of forestal understanding improved and as new influences from European forest practice began to have an impact. And finally, forest policy was continually compromised by other policy agendas and problems, which altered priorities and forced delays. The resulting zigs, zags, and chronic unpredictability of policy only reinforced villagers' irritation and distrust.

From a villager's perspective, therefore, it seemed clear that the new rulers had abandoned the old principles of propriety and fair play and were out to deprive them of valued resources. Moreover, they were inconsistent and unpredictable, and hence richly deserving of distrust and of whatever resistance one could reasonably muster.

* * * * *

In basic structure, then, our story is one of conflict between two parties – reformers in Tokyo and villagers in the hinterland – and of the local and prefectural officials caught between the two. Historical documentation being what it is, however, the story of the former is richly recorded while that of the latter is scattered and scarce. Moreover, the story of the former has a center and, for all its inconsistencies, a detailed narrative coherence whereas the latter is uncentered and incoherent save at the level of broad generalizations. So the villagers' side of our story shows up indirectly, in the reports and complaints of local and prefectural officialdom, in petitions for redress, in the various compromises that Tokyo grudgingly made, and in such crude statistical compilations as police records and reports of damage to the woodland (see, for example, Table 8-1). As told here, the story will unfold primarily as a narrative of Tokyo's performance, followed by some brief "case studies" that suggest how the struggle played out in particular places.

THE BEGINNINGS OF REORGANIZATION: TO BUILD A BETTER FISC, 1871–73

One can plausibly date the start of woodland reorganization to 1871. By then, the new Meiji ruling group had established basic control of the realm, but it also was severely strapped for cash. Leaders saw land sales (and the resulting expansion of their land-tax base) as a way to generate income. But they also recognized their need to establish both compelling claim to any parcels they offered for sale and a set of principles and procedures to guide such sales and maximize their value.[14] These needs led gradually to a realization that the encumbrance of shared-use rights must be repudiated. And the notions of "land ownership" and "market value,"

Table 8-1: Damage to *Kanrin* (by numbers of trees)

Year	Attributed to Artificial Damage			Attributed to Natural Damage
	Arson	Illegal Felling	Total	
1878	48,604	50,511	99,115	–
1879	13,817	–	13,817	–
1880	29,210	71,234	100,444	361
1881	77,201	12,826	90,027	14,675
1882	1,097,111	9,914	1,107,025	333,635
1883	568,028	111,324	679,352	380,952
1884	578,245	231,736	809,981	132,588
1885	489,442	990,673	1,480,115	534,814
1886	959,979	256,340	1,216,319	337,554
1887	1,711,908	92,225	1,804,133	276,378

By the late 1870s the Forest Bureau was well enough organized to collect annual statistics on *kanrin* condition. The figures given here suggest a sharp escalation in the extent of damage during the 1880s.

The meaning of the figues is unclear, however. To some extent the escalation probably reflects improved reportage, thus implying substantially worse damage in earlier years. Also, one can envisage strong reasons why local and prefectural officials would under-report damage, but much weaker reasons to over-report.

Nor is it clear what size trees are involved. One suspects that reports included saplings, perhaps down to about two-three inches basal diameter, but how consistent or complete the reports would be is unclear. And in the case of arson, one suspects that many incidents of intentional burning were attributed to "natural" causes – although areas cut off in the early seventies would have grown back to brush and been much more vulnerable to fire of any sort by the eighties.

(Source: Ringyō hattatsushi chōsakai, comp., *Nihon ringyō hattatsushi: Meiji igō no tenkai katei* Vol. 1 (Tokyo: Rin'yachō, 1960), p. 79.)

---

as propounded by the most powerful foreigners of the day, seemed to provide a mechanism for doing so.

During 1871, this shift in perspective on land control became evident in two ways. One was the adoption in forest-related notices and decrees of the new, universalistic term *kanrin* (later *kan'yūrin*) or "government forest" in place of the diverse Edo-period terms for "lord's forest," each of which carried its own local set of use rights and restrictions.[15] By implication the vast accretion of idiosyncratic use rights that encumbered specific parcels of lord's forest would not automatically apply to *kanrin*.

The other sign of shift in 1871 was acknowledgement of the converse, namely, that government writ would no longer run on non-government land. Specifically, as late as that spring a Cabinet Notice (Dajōkan *tasshi*) reiterated the standard Edo-period prohibition on unauthorized sale and felling of good timber trees of select species (*tomeki* or "reserved trees") regardless of their location, "even at private residences."[16] By that autumn, however, a Notice implicitly acknowledged that government rules no longer applied everywhere by stating that good timber trees, "should be nurtured to the utmost," and even promoted on "private woodlots (*shirin*) by tactful encouragement."[17] And that winter, after forest matters had been assigned to the Finance Ministry, the Cabinet issued a Notice formally repudiating the old policy of controlling disposition of good timber trees on non-government woodlots, stating that henceforth the policy would be limited to government forest.[18]

These changes in woodland policy conformed to the government's broader project of establishing the principle of exclusive property ownership as the basis for a new system of taxation. And in following weeks, as the spring of 1872 advanced, officials in the Finance Ministry moved to facilitate the issuance of land-title deeds (*chiken*). Their goals were to expedite land sales, thereby helping to maximize taxable acreage, while creating a "market value" basis for the new tax system.[19] And that summer, in a Notice intended to facilitate the sale of government woodland – whether well-wooded *sanrin* or scrubby *gen'ya* – for conversion to taxable private property, the Ministry stated explicitly that on parcels thus sold, the long-term rights of others to harvest underbrush would automatically cease.[20]

During 1872, the issuance of titles to arable acreage proceeded apace. Regarding woodland, however, the flurry of notices and instructions was creating problems for the local and provincial officials charged with converting Tokyo rhetoric into hinterland reality. The resulting correspondence revealed to policy makers many of the difficulties inherent in replacing the old use-rights system with ownership. Most notably, some prefectural officials observed that on woodland, unlike arable, one must distinguish between land value and stand value and that they needed instruction in how to do so for determining the value of property.

That particular issue had diverse implications, as the voluminous documentation reveals.[21] There was, to begin with, the matter of "divisible" or "divided" land (*warichi* or *wariyama*), especially those parcels of communal land whose use was periodically shifted from one household to another according to local custom or the ebb and flow of particular household fortune. On such sites, officials asked, were stand and land to be treated separately, or did all belong to the current user, or to the village? If the user, how could the village assure that the property not be mismanaged to the village's detriment? And what about patches of arable or swidden (*yakibata*) that lay within *warichi* boundaries?

Or, in the case of "fixed-term forests" (*nenkiyama*), where the land is understood to be government, village, or household land but where, by contract, the use of stumpage belongs to a different fee-paying individual or group for a fixed number of years – how on this land does one decide "ownership" and its valuation? Or, in the case of "shared-yield forest" (*buwakebayashi* or *bubungi*), meaning government or village land

DATE	*REGULATION NO.*	*KEY RELEVANT POINT*
1871		
11.24	Dajōkan tasshi No.617	Ends *tomeki* policy on private land
1872		
6.15	Ōkurashō *tasshi* No.76	Nullifies *iriai* rights on woodland sold by government
9.4	Ōkurashō *tasshi* No.126	Establishes *kōyūchi* as a land category
1873		
March 25	Dajōkan *fukoku* No.114	Lists land in eight categories
July 20	Dajōkan *fukoku* No.257	Halts woodland sales, temporarily
July 28	Dajōkan *fukoku* No.272	Announces land-tax reform
September 20	Ōkurashō *tasshi* No.134	Explains land-sales halt; calls for "keep" and "sell" categories of land
1874		
March 7	Naimushō tasshi No.1874.9	Broadens *sanrin* to include unstocked acreage
November 7	Dajōkan *fukoku* No.120	Establishes *kan'yū/min'yū* dyad, much expanding acreage to be called *kan'yūchi*
1875		
June 22	CKJ *tasshi* No.3	Relaxes criteria for *iriai* rights on "communal" land
December 24	CKJ *tasshi* No.11	Tightens use-rights criteria for establishing *min'yūchi*
1876		
January 29	CKJ *giteisho*	Final comprehensive statement on disposition of communal lands

Notes: The Gregorian calendar was adopted in January 1873
CKJ = Chiso kaisei jimukyoku

Fig. 8-3. Notable Regulations on Forest Reorganization

on which some individual or group holds a contract to nurture the stumpage, has certain use rights on undergrowth, thinnings, and trimmings, but agrees, as per contract, to share with the landholder the eventual timber yield – how there does one determine ownership and value of the two or more parties' interests?

Finance officials grappled with queries such as these as they flowed in, trying to devise specific answers that were usable and generally consistent. Inevitably, however, variously interpretable and inconsistent formulations emerged and dissimilar practices were sanctioned, which sometimes added more confusion than clarification. As a whole, however, the process gradually made it clear to officials in Tokyo that a large amount of land in and around villages could not be claimed as exclusively government property or sold to individual buyers without creating immense hardship, discontent, and trouble throughout the realm.[22] These exchanges thus forced leaders to confront the whole thorny problem of use rights, rental forestry, and communal property.

The Finance Ministry addressed these issues for the country as a whole in the fall of 1872. In an elaborate twenty-six-clause addendum to an earlier Notice on land titles, it spelled out how to determine ownership of diverse sorts of land in and around villages. It then dealt with ambiguous parcels by introducing the concept of *kōyūchi* ("common-use" or "public" land). The addendum declared, in essence, that all areas which could be measured more or less acceptably and on which ownership could be established were to be so handled. However, parcels where satisfactory surveying was impossible, as well as parcels that were known locally as single- or multi-village woodland and that carried shared-use arrangements, were to be designated *kōyūchi*, with their boundaries indicated as clearly as possible on accompanying maps.[23] The implication of this new formulation seemed to be that, ideally, such areas would in due course become measurable and disposable, ending up as taxable property.[24] But for the time being, existing arrangements there need not be modified.

By then, the process of determining land ownership had slowed substantially, the easy part having largely been completed, and provincial and local authorities were grappling more and more with the contentious issue of woodland control. So, Tokyo's twenty-six-clause addendum elicited a new stream of queries and protests, and further responses by the Finance Ministry.[25] Most pertinently here, in February 1873, Finance officials replied to a query from Akita prefecture by advising that where one or more villages had a clear need to retain control of some land, assessors should establish its value by comparison with similar properties in the vicinity. The village(s) could then acquire the land, receive a title deed, and pay the specified property tax on the basis of the assigned value. The land would then legally be village (*muramochi*) property, not *kōyūchi*.[26] It was a solution that foreshadowed the future, but for villages it meant paying year after year for what had previously been theirs more or less gratis.

With hinterland officialdom frustrated and little progress being made, the government tried again in March to systematize its diverse rulings. Cabinet Decree (*fukoku*) No.114 differentiated land much more elaborately than ever before, specifying eight separate categories.[27] The Decree narrowed the scope of *kōyūchi* somewhat by separating it from "tax-exempt municipal land" (*jozeichi*). And it defined *kōyūchi* to include grassland or other sites on which no tax at all was paid or simply a communal-use tax that local users paid to towns or villages. Such land might be sold and thus converted to taxable "private land" (*shiyūchi*), but only if villagers consented and only with authorization from the Finance Ministry. That Decree may have resolved some of the confusion, but it did so in part by drawing the Ministry more directly into specific local quandaries, which meant that Tokyo continued to receive queries and complaints, mainly regarding the sale of *kōyūchi*.[28]

With that "final" delineation of land categories and criteria, the Finance Ministry felt it was ready to re-invigorate its tax reform effort. In April, it summoned representatives from all the prefectures to Tokyo for a detailed discussion of what the land reform was to entail. Conference sessions revealed the diversity of local land practices, thrashed out a host of ambiguities, and helped participants acquire a sense of shared purpose. Ministry officials then used the feedback from those exchanges to work out the final details of reform procedure.

As of May 1873, then, the government seemed poised to revitalize its land-tax reform. Nevertheless, leaders were still struggling with complications that sprang primarily from the legacy of use rights. They had adopted the makeshift category *kōyūchi* to characterize land whose final disposition they could not yet determine, letting it encompass diverse forms of terrain, parcels carrying shared-use rights, and sites whose boundaries, value, or possession status were ambiguous. How much land was placed – sooner or later – in this *kōyūchi* category is unknown, but it exceeded the total arable acreage of Japan Proper, and it encompassed most of the non-arable that villagers valued most highly.

Officials still seemed strongly committed to shifting as much land as possible, whether woodland or arable, *kōyūchi* or other, to the category of taxable people's property. And they aspired ultimately to scuttle all use rights so as to expedite the shift to a property-based system of taxation and achieve the clearest possible division between government and people's land.

As things stood in the spring of 1873, then, villagers had little reason to rejoice at Tokyo's performance. But still, by adopting the *kōyūchi* category, authorities had in fact compromised on the crucial issue of shared-use rights on much of the usable woodland in and around villages. And they even left it to locals to decide – within limits – which sites qualified as such. As for the rest of Japan's woodland, they seemed prepared to sell it in due course. So while the whole land-reform process was a nuisance and added costs and complications to life, it seemed to be taking the needs of rural folk into account, however grudgingly.

REDIRECTING WOODLAND POLICY: TO BUILD A BIGGER FOREST, 1873–76

During 1871–73, reformers in Tokyo had made considerable progress in identifying landownership on arable acreage, and they had begun selling woodland. They had also, however, discovered many of the complexities they faced in trying to alter the system of woodland control.

During 1873–76, they carried to completion most of the land-tax reform on arable. On woodland, however, they made a basic change in policy direction, and as a consequence their main accomplishments during those years were to work out the implications of that change and develop mechanisms for implementing it.

Tax Reform: A Synopsis, 1873–76

As noted above, by the spring of 1873, the principles, priorities, and procedures of land-tax reform seemed to have been clarified. And, in July, the government issued its elaborate new tax-reform plan, which accompanied the celebrated Cabinet Decree No.272 of the twenty-eighth.[29]

The goals of this reform were to establish, in the European manner, a nationally unified, fully monetized, ownership-based land tax anchored in the property's market value. It was to generate a predictable annual yield, with total receipts initially equaling the estimated total tax yield of the Edo-period produce-tax system. However, because the reformers in fact lacked any adequate basis for establishing the actual market value of most land, whether arable or wooded, the procedures they ended up employing were an ingenious, intricate, and arbitrary set of devices cobbled up for the occasion.[30] The marvel of it all, perhaps, is that these devices actually did work, at least for arable.

The reform process had two basic phases: preparation and certification. During 1873–75, local and prefectural officials used Tokyo's guidelines to prepare provisional listings of ownership, acreage, boundaries, and productive value of taxable properties within their jurisdiction. By the spring of 1875, leaders in Tokyo felt ready to bring the land-tax reform to completion by inspecting and, when satisfied, certifying those listings. To that end, on March 24, Interior Minister Ōkubo Toshimichi took the additional title of Head (*sōsai*) of a new inter-ministerial Office of Tax Reform (Chiso kaisei jimukyoku). A staff of one hundred or more was selected, regional sub-offices were set up, and the new Office was linked to the Finance Ministry by assigning Finance Minister Ōkuma Shigenobu the supplemental title of "Commissioner" (*goyō gakari*) of Tax Reform.[31]

In following weeks and months, inspection teams that included officials from the Office of Tax Reform examined and certified the results of land surveys throughout the realm, doing so on most arable acreage with little difficulty. By the end of 1876, most fields had been evaluated, the tax level set, and the results certified and reported to Tokyo.

In the outcome, it appears, the new tax proved more burdensome to villagers in northeastern Japan than to those in the southwest, evidently because the former's greater vulnerability to weather-related crop failure was not adequately taken into account when establishing the value of farmers' fields.[32] Nevertheless, as a whole, the new system did indeed establish a much more uniform, more completely fungible, nationwide land tax, and it generated roughly the level of income the reformers sought.[33] From the government's perspective, land-tax reform had worked out successfully.

From the viewpoint of villagers, however, matters looked very different. And the reform's implications for woodland usage lay at the heart of village dismay. The processes of surveying, assessing, and determining ownership and tax obligations on woodland were advancing much more slowly, encountering many more complications, and generating far more conflict than on arable. And a major source of this conflict was the government's shifting objectives vis-à-vis such land. These became clear from 1873 onward, and they pitted *kan* against *min* more sharply than ever before.

Tokyo's Shifting Woodland Agenda, 1873

To follow this story, we need to backtrack to one week before the issuance of the tax reform Decree No.272. On July 20, 1873, senior officials abruptly cancelled earlier directives on the sale of government woodland and ordered a halt to all further sales.[34] That cryptic Cabinet Decree No.257 read as follows:

> Herein is announced the cancellation of the notices to all prefectures, dated 1871.8 and 1872.5, in which the Finance Ministry declared its intention to sell to samurai and commoners alike those parcels of desolate and denuded land or *kanrin* that they desire and for which they petition and submit bids to purchase.
>
> However, in unavoidable cases where it is very difficult to stop a sale in progress, a petition for Finance Ministry approval must be submitted, accompanied by a detailed explanation.

Why the government's top leaders had suddenly decided to deprive their regime – as well as samurai whom they no longer wished to support from the treasury – of a source of income was not specified.

Two months later, however, with the new land-tax reform policy announced and reactions from the prefectures flowing in, the Finance Ministry provided hinterland officials with a long statement regarding the halt on sales of woodland. This document, the influential and oft-cited Notice No.134 of September 20, stated:

> On the matter of selling *kanrin* as announced during 1872, such sales did occur, but on more and more sites all the trees were cut off in the name of opening areas to tillage, and the land became denuded. Places

> that in the past only produced trees for construction timber now lack them, and many locations that modulated extremes of wind, rain, cold, and heat and reduced flood and drought are also denuded. This being undesirable, and with such matters in mind, Decree No.257 was issued and the sale of *kanrin* forbidden.
>
> However, henceforth, within *kanrin*, where there are sites [whose sale] will not interfere with fostering water supplies, controlling erosion, or preserving celebrated timber stands, and where there are no other objections to gradual sale, [disposal may be considered. Therefore,] using the following form, examine your jurisdiction without omission and by December 15 of this year report your findings to this office.[35]

The sale of government land was thus not inherently wrong. Rather, it had been handled carelessly, with harmful consequences. Henceforth, the Notice implied, sales could resume, but under new, more rigorous, and more uniform supervision.

It gradually became clear, however, that such supervision would also reflect a changing sense of what government forest was supposed to be and do. In semantic essence, whereas the term *kanrin* had been employed during 1870–72 in place of *ohayashi* and other terms for "lord's forest" – i.e. a simple relabeling of one of the existing official forest categories – from 1873 onward *kanrin* came to be viewed as, ideally, *sanrin* – i.e. the biological category of well-forested woodland, regardless of its earlier official status. In 1874, that new perception found institutional verbalization in the title Sanrinka (later Sanrinkyoku), which identified the government's office of forest administration.

This shift from a primary interest in woodland as saleable commodity to woodland as protective or timber-producing acreage was also evident in the form that accompanied this Notice No.134. It was to be used by officials when reporting the condition of woodland holdings in their jurisdiction.[36] By its choice and arrangement of categories, the form revealed Tokyo's wish to identify as clearly as possible all areas that were unencumbered by old use-rights arrangements. And it revealed that both *kōyūchi* and any other areas ambiguously characterized as carrying shared-use arrangements were to be regarded as basically *kanrin* that the government could choose to retain or sell, depending essentially on its evaluation of legal claims and its assessment of "best use" – whether for timberland, as protection forest, or as taxable private property.[37]

In short, *kanrin* ought to be those parts of Japan's woodland realm that were important for flood-control purposes or that could grow good timber – the classic concerns of protection forestry and production forestry. Only sites that would serve neither of those purposes – indeed, only those sites whose sale would not interfere with water supply, erosion control, or timber development on adjoining areas – were to be sold. How well or poorly this new perception of government woodland would accommodate village needs remained to be seen.

Why the Shift?

A number of factors seem to have been involved in this shift from a policy of valuing woodland primarily for its sales potential to valuing it primarily for its timber potential and flood-control function.

Firstly, by early 1873, Finance officials were thoroughly unhappy with the way their program for establishing the "market value" of taxable private property was developing. It was riddled with confusion, inconsistencies, and probable under-assessments, which prompted them to prepare the much more complex plan of Decree No.272. Because the Decree's implementation would entail a total re-examination of all existing land titles and valuations, the temporary halting of further land sales seemed a sensible preliminary step.

Secondly, Inoue Kaoru, who as a senior figure in the Finance Ministry had been a key advocate of land sales as a source of government income, was no longer in office, having resigned in May.[38] At the time, senior leaders had been locked in bitter dispute over policy toward Korea, and Inoue had opposed an adventurist policy on grounds of government fiscal straits. At the behest of Inoue's rivals, Ōkuma Shigenobu had reworked his figures and concluded that the fisc was relatively secure, hence a militant policy feasible. In protest Inoue resigned. So his departure was accompanied by the creation of financial statistics which suggested that woodland sales were not necessarily of pressing importance to the regime and that sale of the land was, therefore, not necessarily the best policy choice.

Thirdly, diverse domestic voices were protesting sales of one sort or another. Prefectural officials objected to sales that led to reckless logging and consequent erosion and flooding; villagers objected to sales of land on which they had use rights. And perhaps most influentially, naval officials were alarmed by sales that threatened to deprive them of timber for shipbuilding. Their protests halted the sale of forested Mt. Amagi on Izu Peninsula, and, in 1873, their influence may well have played an appreciable role in halting the general policy of selling off government woodland.[39]

In short, there was reason to halt woodland sales at least for the moment. There seemed good reason to question such sales in any case. And a key advocate of the sales had fallen from power. In that situation, the redirection of forest policy became possible. And then a rationale to justify policy change became available when members of the Iwakura Mission began returning from Europe. Key figures on that mission – Ōkubo Toshimichi and Kido Kōin in particular – brought back, primarily from Germany, a richer understanding of the value of forests in government policy. From that perspective, woodland sales seemed hurtful, whereas the formation of a large government-controlled forest realm was highly desirable. And while the role of Ōkubo and Kido in the July decree is unclear, in following months their influence became increasingly pronounced.[40]

Although this gets us ahead of our story, the broader reformist vision that underlay this interest in consolidating government woodland is suggested by a long memorandum that Ōkubo sent to Sanjō Sanetomi, head of the Cabinet, two years later, on May 24, 1875. In it Ōkubo repeatedly invoked the notion of "enlightenment" (*kaimei*) while advocating a wide-spectrum program of social and economic development that would generate exports to pay for imports with which to build a vigorous, new society. He focused on the role of the Interior Ministry, which he headed, in promoting that overall enterprise. And he stated that he was submitting four separate essays in which he would expand specific themes. One essay, he indicated, would dwell on the promotion of forestry, animal husbandry, agriculture, and commerce. A second would treat woodland protection, tree propagation, and the management and finances thereof. A third would discuss organizing and managing the hinterland; and a fourth, the promotion and handling of maritime commerce.[41] It was a policy orientation and an interest in silviculture that would only become stronger as the months passed.[42]

So, in July 1873, woodland sales were halted, and an elaborate new land-tax reform plan was announced.[43] But whereas tax reform had previously been envisaged by its developers as a means of expediting the sale of government woodland, it became a measure that would help to consolidate government control of a vast forest empire by achieving a clear *kan/min* division of the realm. For reformers in Tokyo, that consolidation came to be seen as part of their well-known wish to build a rich and powerful nation (*fukoku kyōhei*). For villagers, however, it seemed to promise only hardship. Consequently the shift in government purpose heightened the difficulty of forest reorganization by spurring further resistance and more woodland damage.

IMPLEMENTING FOREST CONSOLIDATION, 1874–76

The process of creating a large, well-organized and productive government woodland took many years. It did not acquire its full acreage until around 1890 and was not thoroughly operational until years later. The essential character of the project was evident by 1876, however, and during those years, and on through the eighties, it had two basic aspects. One, whose beginnings we can only adumbrate here, was the creation and activation of a government forest administrative system. The other was completion of *kan/min* division, which delineated the areas that government foresters would then administer.

Administering Government Woodland

Almost as soon as the revamped Meiji leadership of 1873 had decided to establish a large, production-oriented government woodland, it began creating the institutions to administer it. In November, it formed a powerful new ministry, the Interior Ministry (Naimushō) with Ōkubo in charge. In the following January, the "Geography Section" (Chiritsukasa), which handled forestry affairs among others, was transferred from the Finance Ministry to the Interior Ministry. By August of 1874, forestry had been separated from such matters as agriculture and riparian engineering and assigned to its own "Forest Department" (Sanrinka) within the Geography Section. Forestry, in short, was no longer to be thought of as a basically fiscal matter, nor even as an aspect of agriculture or village life. It involved *sanrin* and had a purpose of its own. And a government agency now existed that would shape and control that purpose.

For over a year the Forest Department attempted to handle its duties through regular prefectural organs, issuing directives to pursue further surveying, organizing, and regulating of *kanrin* parcels. It even authorized some logging work in select forests. By the spring of 1876, however, with European arrangements better known, authorities moved to establish a direct linkage between Department and woodland. In March the Interior Ministry issued a thirty-one-article protocol (*ketsugi*) that called for a system of regional forest districts (*rinku*) and laid out policy guidelines on forest protection, surveying, record-keeping, rotation harvesting, afforestation, mensuration, and stand-thinning.[44]

During the next few years, extraneous problems slowed the implementation of these guidelines. However, the protocol of March 1876 had made clear the direction of forest policy. And it left little doubt that, insofar as possible, government woodland would in future be handled so as to serve its own purposes, not to accommodate village needs and wants.

As the years passed therafter, that trajectory became more pronounced. In 1879 the much-expanded, rapidly-professionalizing Forest Department was given independent status as the Forest Bureau (Sanrinkyoku). And, two years later, it was shifted from Interior to the newly created Ministry of Agriculture and Commerce (Nōshōmushō), which had been formed explicitly to promote commerce and industry, with both agriculture and forestry seen as parts of that effort. It was a trajectory that left little room for villagers' use rights or subsistence agroforestry.

Pursuing Kan/Min *Division*

Turning to our central topic, from mid-1873 onward, insofar as land-tax reform applied to woodland, it primarily entailed the clarification of boundaries. Doing so enabled reformers to separate *kan* and *min* lands, consolidate government control of timber-producing acreage, and curtail shared-use arrangements there. Thus on March 7, 1874, instructions from Interior to prefectural authorities reminded them that Notice No.134 of the previous September had called for protecting woodlands and timber stands and disposing only of properties whose sale would cause no difficulty in those tasks. The new instructions then expanded the acreage to be protected by stating that one should retain as government woodland not only sites with good stands but also those with the potential to sustain good timberland, "even if they do not contain large trees at present."[45]

As the months passed, Ministry officials responded to a flow of queries seeking clarification on specifics of taxing, selling, and categorizing land. A number of inquiries related to the handling of swidden (*yakibata*) acreage, meaning patches of upland that were cleared, burned over, and used for intermittent, low-grade cropping. During 1873–75, Tokyo's policy gradually clarified and tightened, and, in the outcome, swidden that was found in areas claimed as government woodland, or so situated that its use would impede the management of such areas, came to be defined as government land. Otherwise it could be treated as people's land, its owner granted a title deed and taxed appropriately.[46]

Other, broader issues also came to Tokyo's attention. The sites that generated the most contention were those subject to multiple uses and ambiguity of status, the sorts that had been characterized as "common-use" land (*kōyūchi*) back in the autumn of 1872. Many villagers were most alarmed by the prospect of losing shared-use rights on parcels "privatized" by an entrepreneurial buyer. For village chiefs a common headache was settling title and tax obligations on land carrying multi-village use rights: who was to decide; and why should our village pay the full tax if people from other villages share in the harvest? Rental arrangements continued to pose similar problems.[47]

In a new attempt to sort it all out, on November 7, 1874, the Cabinet issued Decree No.120. It cancelled the eight-category land-classification system of March 1873 (Decree No.114) and arranged all types of land under two super categories, *kan'yūchi* and *min'yūchi*, thus formally establishing this basic dyadic perception of woodland. The common-use (*kōyūchi*) and tax-exempt (*jozeichi*) categories of No.114 were to be divided between the two, and in essence, all parcels for which individual or village ownership could not be established were to be incorporated in the *kan'yūchi* category.[48]

This Decree No.120 identified government land as consisting of property of the Imperial Household and its Ise shrine, land owned by prefectural governments, and areas claimed for official use by offices of central and regional government, the judicial and police systems, and the army and navy, plus a broad miscellany of properties – roads, waterways, cemeteries, etc. – for which no taxable ownership was to be established. And the informal term *kanrin*, which initially had been used in lieu of *ohayashi* to embrace all government woodland, came to denote only those parcels that the Forest Department administered directly, as distinct from acreage, wooded or not, that was administered by the Imperial Household, prefectures, or other government agencies.

As for people's land, the Decree declared that it could be freely bought and sold, but that any major change in land use must be approved by the Interior Ministry. That proviso reflected the concern with erosion, flooding, site deterioration, and the impact of change on any adjoining government land. The old, universal Tokugawa policy of "reserved trees" (*tomeki*) was dead and gone, but a new, more vague, and more all-encompassing policy of government oversight of people's land based on environmental criteria – a policy that would flourish in subsequent decades – had been enunciated.

During 1875–76, as tax reform on arable acreage moved to completion, instructions from Tokyo continued to address local problems relating to *kan/min* division. A critical issue was the question of what "evidence" proved the "ownership" that Decree No.120 required.[49] In the outcome, two criteria became decisive for settling ambiguous cases. For parcels to qualify as people's land, claimants must show either evidence of prior tax-paying use rights or evidence of past stand-nurturing work.

The problem posed by this formulation was that for centuries villagers throughout the realm had customarily harvested natural growth as a shared-use right, paying no tax and simply leaving the sites to regrow or reseed their own grasses, brush, and trees. By these new rules all such areas – which by earlier, less restrictive criteria could have been designated *min'yūchi* – were to become *kan'yūchi*. So these guidelines enabled the government to lay claim to vast acreages of woodland – millions of hectares by an informed estimate of the 1890s – that might have value as productive timberland regardless of their former actual usage.[50]

Unsurprisingly, insistence on these criteria only generated more queries and complaints. Among the scores of replies and rulings that Tokyo sent out to prefectural officials during 1875–76, two of June 1875 merit particular note.

On the twenty-eighth, in the less portentous of the two, Ōkubo responded to the needs of prefectural officialdom by notifying governors that they could dispose of parcels of government woodland under their own jurisdiction (*kan'yū rin'ya*), but not parcels of *kanrin*, which came under the authority of the Forest Department.[51] While that notice did assure regional officials some flexibility in handling their own resources, it also had a restrictive effect. It explicitly affirmed Tokyo's right to control directly any woodland that it designated *kanrin* – most of which had been, until recently, lord's forest under daimyo control or shared-use land of ill-defined local control, and more of which Tokyo intended to sequester in future.

A week earlier, in a ruling that generated far more complications, Ōkubo had responded to prefectural queries by confronting yet again the stubborn problem of use rights and the widespread fear of losing them. In an effort to allay distrust and dissuade villagers from hurriedly felling trees, removing other goods, or burning sites to make them seem undesirable as *kanrin*, Ōkubo's Office of Tax Reform issued Notice No.3, which assured them that existing use rights would be recognized in the communal lands of villages and local associations (*mura kōyūchi* and *kumiai kōyūchi*), even when the documentation of such rights was slender.[52]

Ōkubo's intent was to address an important source of resistance to woodland reform. However, the ruling proved mischievous because many villagers misconstrued its implications to their subsequent regret.

To explain, as noted above, Decree No.120 of November 1874 had established that parcels lacking clear ownership, including those erstwhile labeled *kōyūchi*, were to be considered one or another type of government land. And subsequent decisions had all reinforced this understanding, causing much acreage on which villagers had long-held use rights to be so designated. Because of the ambiguity that had long marked the relationship between government, village, and communal land, in following months and years many villagers came to perceive this Notice No.3 as securing their use rights on "public" or "common-use" lands in general, meaning any land without an identified owner. On that presumption some chose to cede parcels of woodland to the government as *kan'yūchi*, rather than to claim them as *min'yūchi* in the too-cunning belief that such a choice would enable them to continue enjoying customary use rights while evading the new taxes on "private" property. And the fact that use rights on government as well as village land did indeed continue to be widely tolerated – some with and some without Tokyo's consent – seemed to validate that view.

However, during the eighties (to look ahead briefly) more and more villagers were disabused of that belief. In pursuit of the government's timber-production and water-control objectives, forestry officials step by step tightened their control of *kanrin* usage, imposing fees and adding restrictions and red tape. In consequence more and more villagers discovered that the tradeoffs in using "public lands" were growing ever less advantageous. The wide-ranging assault on use rights struck them as further evidence of Tokyo's perfidy, as betrayal of assurances given in this Notice No.3 and numerous other rulings. So villagers did what they could, resorting to yet more reckless felling and burning (see Table 1) in hopes that once the land had been ravaged, the government would decide to dispose of it as unfit for use as *kanrin*.[53]

In other cases, Ōkubo's Notice No.3 notwithstanding, distrustful villagers continued during 1875 to suspect that, no matter how it worked out, *kan/min* division would deprive them of use rights. So they kept on felling trees as speedily as they could before property rulings were finalized.

Prefectural officials continued to struggle with the problem of reckless felling. Some governors called on Tokyo to allow more flexibility in designating people's land; others fretted that too much looseness would only lead to more denuding and damage. But Tokyo's attempts to accommodate these divergent views just added new inconsistencies to policy.[54] So, in what proved to be the government's last general set of instructions, on January 29, 1876, the Office of Tax Reform issued a six-clause directive intended to guide the sorting out of communal land.[55]

The gist of this directive followed that of the Cabinet's land-classification Decree No.120 of November 1874: where evidence of established practices of stand nurturing or formal Tokugawa recognition of village possession was lacking, contested parcels of common-use land were to be designated government land. However, the attention the directive gave to distinguishing between categories of land carrying shared-use rights (multi-village, single-village, and *kumiai iriaichi*) suggests how the processes of *kan/min* division, land sales, and tightening government claims to *kanrin* were generating tension not only between regime and populace but among local officials and villagers themselves as the specifics of old use-rights arrangements came under attack from several directions.

As of 1876, that is to say, Tokyo reformers evidently were satisfied – at the level of countrywide rulings, anyway – with the regulations they had already issued. In their eyes, their policy positions were clear, consistent, and appropriate. Villagers, however, saw things otherwise, as the next dozen years would reveal. And hints of this persistent rural discontent and of the factors that promoted it are evident in the experience of select localities.

SOME CASE STUDIES

As the above narrative suggests, woodland-related actions of the Meiji government are amply recorded. The actions of prefectural officialdom are also well documented, but those of local officials and villagers are much less fully preserved. Those last mostly addressed particular local concerns, so brief summaries of discrete cases may illustrate them, as well as the diversity of conditions and issues involved in woodland reorganization.

Scholarly studies give us glimpses of woodland reform as it unfolded in several locations. Within Japan Proper the forest areas most highly prized by reformers, and hence the woodland disputes of most concern to them, were situated in central Japan. However, non-arable in even the far reaches of the archipelago also became embroiled in the ruler/ruled contest for resources. So let us look at three disputes in the outer areas – Tsugaru, Shōnai, and Miyazaki – and then a pair – Ise and Kiso – in central Honshu, giving fullest attention to the richly-documented case of Kiso.

Tsugaru

In the far north of Honshu, in former Tsugaru domain, sand dunes lined the Sea of Japan coast for some 25 km south of Lake Jūsan. The dunes were delicate terrain that shielded rear areas from the sea, and they had long been managed as lord's forest, with locals enjoying designated use rights. With dissolution of the daimyo domains in 1871, the old controls grew lax, afforestation work was abandoned, and in following years parcels were offered for sale in line with Inoue Kaoru's program of land disposal.

Land sales appear to have continued despite Tokyo's 1873 shift in policy direction, but finally in January 1876 the Office of Tax Reform's six-clause directive prodded prefectural officials in Aomori to designate the entire dunes area as government woodland. That action halted all further sales and threatened residual use rights. This reversal of policy spurred locals to begin felling trees aggressively before authorities could stop them.

The dunes were clearly not the sort of mountainous *sanrin* that Tokyo seemed to want, so, in 1880, with the area partially denuded, local officials representing sixty-six villages petitioned the prefecture to transfer them to the villages without fee. In response the Forest Bureau asserted firmer control. Dissatisfaction persisted, and two more petitions and nine years passed before forestry officials in Tokyo and Aomori agreed to a long-term rental arrangement. Officially the land remained *kanrin*, but any trees would belong to the villagers who planted and nurtured them, doing so in accord with government regulations.[56] In essence, a shared-yield forest (*buwakebayashi*) system had been re-established under Forest Bureau supervision, and in following years the dunes regained some of their forest cover.

Shōnai

Some 240 km south of Lake Jūsan, in former Shōnai domain, sand dunes lined the coast for another 34 km north and south of the Mogami river estuary. There too the area had been handled before 1868 as lord's forest with regulated local usage.[57]

During the early seventies, local villagers gradually expanded their use of the dunes. Then in 1876 the six-clause directive, with its call for clear *kan/min* division using the restrictive criteria of stand nurturing and documented use rights, prodded prefectural officials to commence a formal demarcation of boundaries. The surveyors first focused on arable and residential land, finishing that task by November. During 1877, they worked on woodland, completing the job in a rough and hurried manner that relied heavily on "eyeballing" sites and following existing maps. In the outcome, the government reclaimed much of the dunes area as erosion-control *kanrin*, but neither forestry officials nor locals were satisfied with the outcome.

Even before the initial survey of 1876–77 had been completed, locals began petitioning to have some of the areas returned to them gratis and other areas reopened to their use as shared-yield stands. During the eighties, however, when officials examined the petitions and resurveyed the dunes, their judgements generally worked to local disadvantage. And over the next twenty years, as officials gradually made decisions on their backlog of petitions, nearly all the efforts to reverse *kanrin* rulings were formally rejected. After 1889 villagers turned to formal lawsuits, which the new Meiji Constitution made possible, but most of those also were denied, except for some shared-yield proposals that meshed with the forest-development plans of the Forest Bureau.

In the end, then, in Shōnai, as in Tsugaru, *kan/min* division was completed. Most of the dunes remained *kanrin*, but on them various concessions were made that allowed local usage of one sort or another to continue. The ruler-ruled tensions had thus been meliorated, but mainly by falling back on modified versions of older, shared-yield arrangements

Miyazaki

Far to the south, in Miyazaki prefecture in southeast Kyushu, prefectural officials began to implement their *kan/min* division at the end of 1874. As survey results were recorded during 1875, officials struggled to apply the required criteria of documented usage and evidence of nurturing. The project was delayed during a redrawing of prefectural boundaries in 1876 and then by turmoil associated with the Satsuma Rebellion of early 1877. By September 1879, however, the surveying was completed, and officials forwarded the required maps and land registers to Tokyo.

Even as areas in Miyazaki were being designated *kanrin*, locals moved to cut trees down before authorities could prevent it. And during the eighties government personnel conducted further surveys to clarify vague and disputed boundaries, to evaluate timber stands and sites, and to sell unwanted parcels. In some of those resurveys, the authorities claimed as *kanrin* parcels that previously had been designated people's land, and in such cases villagers not only lost land but realized that they had been paying a tax on property not their own.

The scale of such matters might be miniscule, but for villagers eking out a subsistence-level existence, small matters could loom large. And there were, after all, principles of fairness and propriety involved, as suggested by this petition for a refund of several years' worth of taxes that totaled about five *sen* (roughly five cents, or perhaps a couple of American dollars in today's money).[58]

Petition Regarding a Change in Land Tax

Hyūga-kuni, Minami Naka-gun Sakida-mura, Ōaza Nagata No.3234
on a site of 9 *se* (1/11 ha.)

3 *se* (1/30 ha.) of hillside, taxed at 6 *rin*

Although this area had previously been *kanrin*, at the time of the land tax reform in 1879 it was mistakenly categorized as people's land. Now, as this has been discovered, although regrettably belatedly, [the parcel] has been restored as government land. So I wish the tax levy refunded. Certainly, while [the parcel] was in my name, in my use of the land there was no felling of trees or such like. So do I now petition.

10 July 1886 Minami Naka-gun, Sakida-mura
Kōno Kesachi [?]

[To]
Governor of Miyazaki-ken, Tanabe Terumi

The above is certified to be as stated and without discrepancy.
Headman: Kōno Sansō [?]

The headman forwarded Kōno's petition to Governor Tanabe, who responded favorably, instructing local officials to settle the matter appropriately. So, although the final act is not reported, perhaps Kōno did recover his several years' worth of 6 *rin* in annual taxes.

Not everyone settled for pennies when resurveys of the 1880s shifted more Miyazaki woodland into government hands. Villagers still needed *gen'ya* goods, so arson and theft continued to occur. During the nineties, prefectural leaders urged the Ministry of Agriculture and Commerce, which then housed the Forest Bureau, to convert large areas of *kanrin* to people's woodland. Tokyo resisted, however, calling for further investigation and suggesting more use of shared-yield stands. Gradually, as the Meiji era neared its end, such shared-yield contracts were negotiated, while other *kanrin* parcels were converted, some becoming village land,

and some private land. As those settlements were arranged, the level of protest diminished.[59]

Ise

Tokyo moved to control woodland in central Japan sooner than it did in peripheral regions, so village resistance appeared there that much earlier. In Watarai (today's Mie) prefecture, destructive disputes flared in the mountains around Ise shrine from the early seventies onward.[60] In the legal arena, the contest was fought in terms of two entangled agendas. Tokyo and prefectural officials dueled primarily over the outcome of *kan/min* division. Villagers struggled mainly to retain use rights essential to their livelihood, regardless of the land's legal classification. At a cruder level the battle entailed arson, theft, and attempts at police control.

The forests near Ise shrine had long been utilized by locals as shared-use land, especially for the production of charcoal and firewood. But when the Meiji government laid claim to all shrine and temple woodland in early 1871, the area became *kanrin*.[61] That change was welcome to shrine officials, who for several years had been alarmed by the extent of deforestation that charcoal makers were causing. But locals, who relied heavily on kiln output, were concerned by the threat to their livelihood. They complained, and that autumn Watarai authorities reaffirmed charcoalers' right to harvest wood.

After Inoue Kaoru authorized the sale of forested land in 1872, the locals submitted petitions to acquire the areas. When Finance officials finally responded in June 1873, however, they only agreed to sell parcels to the highest bidder, not to the locals who had previously held use rights there. And they would authorize those sales only after receiving satisfactory information on acreage and stand composition. That offer, unsatisfactory as it was, proved shortlived because a month later, as previously noted, all sales of government woodland were halted while Tokyo leaders geared up for general land-tax reform.

So the woodlands of Ise remained *kanrin*. In January 1874, Watarai officials launched their first effort to survey them to determine "keep" and "sell" parcels, as per Notice No.134 of the previous September. But in December 1874, the Interior Ministry informed them, in line with the earlier-noted Cabinet Decree No.120 of November 7, that the Shrine mountains were, in any case, to remain *kanrin*. Other adjacent woodland that villagers had been using was also designated *kanrin* because of its earlier status as lord's forest. This development posed such a severe threat to residents of nearby intermontane villages that in July 1875 the governor of Watarai petitioned Tokyo for the conversion of designated areas to people's land.

Meiji leaders had no intention that land close to the premier imperial shrine should be converted to people's land. But if imperial benevolence could not be manifested at Ise, after all, where could a loyal subject expect to enjoy it? Moreover, should a major wildfire erupt there – whether due to arson or accident – it could prove disastrous.

In essence, the repudiated concept of use rights provided Tokyo reformers with a solution to their dilemma. Even as prefectural officials were trying to implement *kan/min* division, villagers repeatedly petitioned Tokyo to recognize their right to cut brushwood (*zōki*) for charcoal production. In April 1874, the Interior Ministry did allow it, but only in cases of special need. That concession hardly sufficed, so villagers submitted more appeals, and, in March 1875, the Interior Ministry authorized, for a set fee, a precisely-specified amount of cutting by people from identified villages, appending to the permit a strict prohibition on cutting any "good timber trees."[62]

A year later, Ise villagers petitioned again to cut fuel wood, and again the Ministry approved. But it also tightened its procedures, instructing the villagers to adopt a system of rotation cutting. In subsequent years, Tokyo's oversight tightened further, limiting the cut so as to aid forest recovery, and doing so despite local protests that they could not survive without producing more charcoal.

Facing unacceptable restraints on their use of *kanrin*, some villagers evidently tried to reduce areas to such worthlessness that Tokyo would dispose of them. Beginning in the spring of 1876, the time of year when food reserves were lowest and village life most difficult, fires of suspicious origin began erupting on the lower hillsides, along with instances of wood theft. Particularly large wildfires in March 1880 led to frantic efforts by local and prefectural officials to get restrictions eased, and after some dithering Tokyo approved more extensive harvesting. That concession quieted the unrest and halted the arson, at least for a few years. But in Ise, as elsewhere, the basic tension persisted, easing slowly only as other socioeconomic changes gradually altered local circumstances.

Kiso

The most well known and most richly-documented instance of forest disputation sprang from *kan/min* division and restriction of use rights in the Kiso River valley. The area is unrepresentative in a couple of ways. It may have held the finest stand of timber in Japan Proper. Also it was one of the first major forest regions to come into the hands of the new Meiji rulers, which meant that local protests, petitions, and illicit activity erupted sooner there than in most areas. And later, during the 1880s, Kiso came to be seen by those rulers as the most treasured piece of a glorious new Imperial Household forestland (*goryōrin*) that would center on the sacred mountain, Fujisan, thereby channeling the godly powers into permanent support for the Household, the regime, and the realm.

Whether prized for such a lofty reason, or simply as high-grade forest of great value to the regime, Kiso became the woodland area that Meiji leaders were most determined to retain, regardless of the legitimacy of local claims to use rights or the local hardship their denial caused.[63] The most celebrated of the Edo period's three "beautiful forests" (*birin*), Kiso constituted a large, contiguous area of clearly-identified lord's forest (of Owari domain). Moreover, thanks to topography, climate, and a direct river route to Nagoya, it was a nearly-ideal site for the growth of *sugi* and *hinoki*, which were precisely the trees most desired by the rulers.[64] Finally, Kiso's dendritic system of narrow valleys supported little agriculture and therefore a relatively-small resident population for Tokyo to confront.

By the summer of 1873, when Tokyo re-energized its program of land-tax reform and *kan/min* division, Kiso villagers were already in conflict with prefectural authorities. As Tokyo had stipulated in 1871, the latter viewed all the former lord's forest as *kanrin*, and accordingly they denied the long-recognized use rights of villagers. When angry and needy locals persisted in removing wood in defiance of the new prohibition, officials deployed patrols and began arresting and prosecuting those whom they caught with illicit goods.

At the time Kiso was part of the recently created Chikuma prefecture, and its governor was the vigorous Motoyama Moritoku, whose energetic pursuit of *kan/min* division and *kanrin* consolidation won him lasting and rather unfair notoriety. In line with Notice No.134 of September 1873 and the Interior Ministry's supplemental instructions of March 7, 1874, Motoyama set men to work surveying his realm. They prepared the required maps and land registers, attempted to organize material in the designated *kan'yū, min'yū*, and *kōyūchi* categories, and tried to assure that all areas capable of producing good forest would ultimately become *kanrin*. Although unclear, it is probable that prefectural surveyors began their work soon after snow started melting in spring 1874, and they had finished by July 18 of that year.[65]

During the Edo period, essentially all of the valley, except for the small parcels of arable and *gen'ya* around the handful of streamside villages where farmer/loggers lived, was lord's forest. Much of the accessible woodland had been categorized as "open forest" (*akeyama*), meaning areas on which locals had well-established rights to gather fuel wood. Generally they could harvest for a set fee or in some places without charge. Whatever the particular arrangements, however, all parties agreed that open forest was lord's forest. Therefore the surveyors of 1874 recorded much of it as *kōyūchi*, as per Tokyo's rulings of 1872–73.

In July, officials from the Interior Ministry traveled to Kiso, examined Motoyama's results to date, and found them satisfactory. With so little land identified as people's land, the emerging picture was of a Kiso realm divided between government land and *kōyūchi*. The former was the closed portion of Owari lord's forest, and the latter the extensive open forest with its local use rights. By early September, village leaders had approved this *kan'yū/kōyū* division in the new registers, and on November 17, 1874, the prefecture certified the results to higher authority. [66]

It appears that locals saw this outcome as one that gave the government its *kanrin* while leaving them with their use rights on *kōyūchi*.[67] However, even as the survey was being finalized in Kiso, the Cabinet in Tokyo had issued its earlier-noted Decree No.120 of November 7, with its new guidelines that called for the division of all *kōyūchi* into *kan* and *min* categories. So the 1874 survey would have to be redone. After receiving these guidelines, Chikuma officials proceeded to do so, probably starting the job when the weather had warmed up and travel become easier in early 1875. They assigned to the "people's" category both those sites where documentation showed that villagers had planted the trees and all areas that villages had long used as shared-use land for daily necessities (as distinct from saleable harvest). In their report they cited the Office of Tax Reform's Notice No.3 of June 1875 as authorization for doing so. In July, they asked the Interior Ministry to approve their action on replanted sites, and two months later on the rest, the shared-use land.

The latter request proved acceptable, and the Ministry approved it on September 27, about two weeks after receiving it. The former request, however, presented a problem. On one hand it obviously satisfied the criterion of "stand-nurturing work" that allowed *kōyūchi* to be classified as people's land. But on the other, it implicitly involved land that would grow good timber, which meant it should be retained as government land. Wanting to retain it but lacking solid justification, the Interior Ministry took no action.[68]

Then matters were redefined. On December 24, in an apparent effort to reduce the amount of woodland being shifted to the *min'yū* category on the basis of stand-nurturing, the Office of Tax Reform ruled (in Notice No.11) that all petitions using such a rationale must provide explicit documentation of particular claims.[69] In Kiso, as elsewhere, that requirement undercut many claims because records had long since disappeared or because planting and aftercare had been done with no record keeping. With all the Kiso woodland in question recognized as former lord's forest of one sort or another, and hence *kanrin*, that ruling, together with the six-clause directive of January 1876, undermined both the Chikuma survey of 1874 and the categorization effort of 1875.

By the spring of 1876, Tokyo reformers had both a sufficiently clear sense of their woodland objectives and a sufficiently well-developed forest administrative system so that they were ready to address the Kiso matter. In April, the Office of Tax Reform abruptly cancelled its earlier approval of Chikuma's proposal on shared-use acreage and repudiated the prefecture's entire effort at division of *kōyūchi*. It reminded prefectural officials that all the areas in question were indeed lord's forest. It asserted that villagers had rights of entry only into areas lacking good timber stands and that all stands must be evaluated before decisions about assignment can be made. Finally, it said, after such evaluations are completed – and in full accord with the Office's most recent instructions (of December 1875 and January 1876) – the petition for approval must be resubmitted.

Chikuma officials, now tightly wedged between Tokyo reformers and Kiso residents, responded promptly. On May 4 they argued, in defense of their requested division, that villagers burned grassy areas, nurtured new growth, and planted the areas in question to pine, chestnut, horse chestnut, and oak (*nara*) – trees of use to themselves but not as timber. They assured Tokyo that villagers would safeguard valued stands of timber trees (*goboku*), and offered to discuss the matter further.[70]

At the time, the Interior Ministry was getting on with its plans to survey and organize *kanrin*. But here in Kiso, the anticipated jewel of its woodland realm, the basic coherence of its forest seemed to be in fundamental jeopardy. Accordingly – and even before Chikuma had prepared its response to the April directive – the Ministry despatched an investigating team whose lead members reached Fukushima in Kiso on April 28. Then the Ministry received Chikuma's May 4 reply and rejected it harshly on the twenty-seventh. Asserting that Chikuma's survey had been faulty, that later changes in boundaries had been made, that indeed there was doubt whether any real survey had even been made, it announced that more investigators were being despatched.[71]

During the summer, investigating teams composed of officials from the Ministry, prefecture, and locality methodically worked their way across Kiso. They examined boundaries, installed and numbered marker posts, marked boundary trees, measured distances between them, used old records to check woodland stand composition, estimated acreages, and prepared maps and registers that recorded all this information. As surveyors completed their work in successive villages, they had local officials certify the results, and by August the task was completed.[72]

At that point, the whole project appears to have come to a standstill, probably because a cluster of problems led Tokyo in August 1876 to dissolve Chikuma and assign the Kiso region to Nagano prefecture. Nagano officials then spent several months obtaining and digesting the pertinent Chikuma records, reorganizing their own internal administration to absorb their large new territory, appointing and despatching subordinate officials, and working out lines of communication with districts, towns, and villages.

Finally, on June 18, 1877, Nagano officials moved to implement the new survey findings. They requested that the parcels of *kōyūchi* to be reassigned as people's land be so designated without fee and that sale of timber stumpage (*goboku*) on such land be approved as proposed. The request, however, pertained to only 30,000 out of some 220,000 *chōbu* of open forest, which meant that Kiso villagers were about to suffer an 85 percent defeat. Nevertheless, officials within the Interior Ministry reportedly disagreed among themselves as to how such matters were supposed to be handled. Not until June 1880 was the request finally approved, and with it came a firm requirement that any logging be handled in a way that did no harm to *kanrin*.[73]

In its totality, this 1877 survey in Kiso assigned about 38,000 *chōbu* of the region to villagers as people's land while nearly 350,000 was held as government land.[74] And while various use rights were recognized on parts of the latter, they were increasingly burdened with fees and restrictions. The outcome was not at all satisfactory to locals, whose welfare was at serious risk, and in May 1881, twenty-one villages in the valley prepared a joint petition calling for a resurvey of the region.[75] That petition was but the first salvo in what became the so-called "Kiso Incident" (Kiso *jiken*), a struggle for rights of forest use that dragged on for decades.[76]

Nor was the government satisfied with the way things were working out in Kiso. As noted earlier, the Interior Ministry had gradually reorganized and enlarged its central forestry organization. It had also set up branch offices and placed one of them in Kiso. In July 1880, officials there notified the Forest Bureau that some boundary markers were rotten, that others had been moved, and that people were plundering valued timber in areas previously designated *kōyūchi*. Indeed, in some spots they were hacking everything down so that sites would be judged unfit as forest and converted to people's land.[77]

In response to the news that Kiso *kanrin* was still under siege despite the supposed completion of *kan/min* division there, Tokyo officials ordered new, more careful surveys and boundary marking. The surveyors were informed that their primary concern was to deal with those boundaries that were still unclear or incorrectly recorded, or where markers had rotted, been destroyed, or moved by locals, and to check areas where trees had been felled illegally. Tokyo's concerns clearly were not those of the twenty-one village petitioners, and although some resurveying was done in Kiso in late 1881, the results were hardly satisfactory to those petitioners. Both prefectural and local spokesmen lodged strong objections to the way the survey was being handled, but it was resumed in April 1882 after the snow melted.

Unsurprisingly, this resurvey only sharpened the tension between Forest Bureau officials and Kiso villagers, contributing to the long-drawn-out Kiso Incident. Although that affair lies beyond the scope of our present inquiry, the process recounted here, which precipitated it, does suggest how the basic conflict of interests between Tokyo and villagers over woodland usage was brought into focus by the 1870s *kan/min* division and the government's creation of a woodland empire managed to meet its own needs.

EPILOGUE: THE COMPLETION AND AFTERLIFE OF *KAN/MIN* DIVISION

As noted at the outset, the story of *kan/min* division had no real end point. It just petered out. But even that process had stages, and it involved some interesting twists and ironic turns. The basic process of separating government and people's land, identifying *kanrin*, and establishing government control of it was completed by 1889. Then a long period of organizing and sorting out – which entailed a modest reversal of earlier policy – continued into the 1930s. And finally, from the 1950s onward the whole issue became irrelevant for reasons that remind us of just how thoroughly extraneous factors can determine the flow of history, rendering moot, even trivializing, our shrewdest maneuvers, keenest calculations, and most diligent labors.

The Completion of Kan/Min *Division, 1876–89*

As the experiences of Kiso and other localities have shown, *kan/min* division was far from finished in 1876. Indeed, the most contentious years lay ahead. However, the essential logic and character of government policy and of the conflicts it engendered were visible by that year, so suffice it here to summarize subsequent developments.

From 1876 to the early eighties, the pace of woodland reform eased substantially, primarily because of civil unrest, most famously the Satsuma Rebellion. These problems diverted energies, depleted the Treasury, and prompted Meiji leaders to avoid actions that would further inflame the hinterland. During the eighties, however, with rebellion suppressed and *kanrin* forest administrative structure substantially more developed, Tokyo returned to aggressive surveying and boundary marking. Step by step it restricted and complicated the exercise of use rights on *kanrin,* and it promoted logging and reforestation there in what developed into a long-term program intended to transform ever more of the archipelago's deteriorated woodland into *sanrin.*

During the early eighties, this revitalized effort, together with the hardships inflicted on poorer villagers by the deflationary retrenchment policies of Finance Minister Matsukata Masayoshi, produced a surge of reckless logging, woodcutting, and arson. An easing of retrenchment in 1884 reduced the latter source of discontent, but during the next five years Tokyo's teams of property and boundary inspectors assigned vast amounts of indeterminate land to the *kan'yūchi* category (see Table 8-2), which angered villagers and sustained the assault on woodland. To counter these acts and secure its forests, the government step by step restricted villagers' use rights and employed its growing staff of forestry officials and police to catch and prosecute those charged with vandalism and theft.

By 1890, *kan/min* division was essentially complete; Tokyo had secured the woodland it sought. On the islands south of Hokkaido, the

Table 8-2: Expansion of *Kanrin* (by numbers of sites and acreage: *chō*)

Year	No. of Sites	Total *Kanrin* Acreage (*chō*)
1880	82,795	5,259,182
1881	82,896	5,363,552
1882	82,983	5,388,787
1883	86,275	5,500,938
1884	88,297	5,576,595
1885	93,138	5,635,257
1886	113,018	6,893,880
1887	130,978	7,010,836
1888	142,795	7,691,671

(Source: Ringyō hattatsushi chōsakai, comp., *Nihon ringyō hattatsushi: Meiji igō no tenkai katei* Vol. 1 (Tokyo: Rin'yachō, 1960), p. 61.)

Forest Bureau controlled some 7.7 million *chōbu* of *kanrin*, while prefectural governors and other government agencies controlled another half million or more as *kan'yū rin'ya*. The Imperial Household held 1.6 million *chōbu* as *goryōrin*, and some 8.4 million were registered as *min'yūrin* of one sort or another. The portion of Japan Proper's woodland classified as government forest thus stood at 53 percent of the total, having risen from 31 percent in 1880.[78] And while use rights of diverse sorts still survived on many parcels, they were subject to severe restrictions, obligations, and fees that sharply reduced their utility to most villagers.

The Afterlife of Kan/Min *Division, 1890–1930s*

Almost as soon as its woodland empire was formed, the Forest Bureau found itself under pressure to reduce it. The creation of an elected National Diet in 1890 gave the rural populace a far greater voice in affairs than hitherto, and under pressure from rural Diet members, the government began a gradual program of disposing of woodland. In that process, the Forest Bureau employed, as much as possible, its own "best use" criteria, retaining timberland and critical water-control acreage, selling or donating only marginal areas to entrepreneurs, villagers, cities, towns, villages, fanes, and other entities. Foresters regarded about 10 percent of the 1890 *kanrin* acreage as *gen'ya*, and by 1920 the Bureau had disposed of most of it (some 750,000 *chōbu*), using the income to pay for reforestation and rehabilitation projects on the land it retained.

By then entrepreneurial lumbermen were a substantial presence in forest operations, both as landowners and as leaseholders on *kanrin*. They and village landlords had been major beneficiaries of the woodland-sale program, which had sold parcels to the highest bidders, hence those with the most money to invest. For most villagers, therefore, the sale of woodland did not restore old use rights. So discontent remained, and problems of arson and theft persisted, though at rates much reduced from that of the 1880s.

Rendering History Irrelevant, 1950-Present

The later-Meiji pattern of woodland organization in Japan lasted through World War II. Then, in 1947, the several types of government-controlled land (including *goryōrin*) were consolidated under a reorganized and retitled Forest Agency. Its new name, Rin'yachō, acknowledged that during preceding decades government foresters had come to play a substantial role in both *sanrin* and *gen'ya*, as administrators of government land and advisors and assistants on people's land. However, the basic legal division between government and people's land survived, and the earlier patterns of woodland use also persisted, with government and entrepreneurial lumbermen wanting *sanrin* and its yield of logs while most rural folk wanted *gen'ya* with its harvest of fuelwood, fodder, fertilizer materials, foodstuffs, and mulberry leaves for silk-production.

Beginnning in the 1950s, however, everything changed. Fossil fuels gradually replaced firewood and charcoal; commercial fertilizers (made largely from fossil fuel) replaced green fertilizers; gasoline-powered machinery replaced fodder-eating draft animals; artifical fibers (mostly from fossil fuels) replaced silk and its mulberry-munching worms; and the urbanizing population lost interest in most of the foodstuffs that *gen'ya* could provide. So subsistence agroforestry petered out and within decades *gen'ya* went from necessity to irrelevance, abandoned and left to grow as it chose.

Meanwhile, from the 1960s onward, Japan became an active participant in the global timber trade. The loggers and lumbermen who worked Japan's mountainous terrain found themselves utterly unable to compete with producers abroad, and the level of the domestic harvest steadily dropped. As logging lost its profitability for both entrepreneur and Forest Agency, reforestation and aftercare became uneconomic, an investment that carried no promise of eventual return.

Instead, the grand coniferous forests of Japan's sprawling *sanrin* mostly just stood there – and stand there today – aging, suffering from windthrow and snow breakage, slowly losing their residual economic value. So for government, as for villager, the woodland of Japan has ceased to be a resource worth fighting for. The underlying logic of *kan/min* division and its associated struggles has been – for the moment at least – utterly destroyed. What the early Meiji quarrel was all about, few people today could really say.

APPENDICES

Appendix A

Measurements and Glossary

MEASUREMENTS

chōbu	measure of area (0.99 hectares; 2.45 acres)
kan	measure of weight (3.75 kg; 8.72 lb)
ken	measure of length (1.82 m; 1.99 yards)
koku (grain)	measure of volume (180.51 liters; 5.12 U.S. bushels)
koku (wood)	cubic measure (1x1x10 shaku; 0.27 m^3; 9.70 $ft.^3$)
sen	a coin; one one-hundredth of a yen
shaku	measure of length (30.3 cm; 0.99 ft.)
sun	measure of length (3.03 cm; 1.2 inch)
yen	currency: in 1875 = approx. $1.00 US in 1935 = approx. 25 US cents in 1970 = approx. 0.28 US cents in 2005 = approx. 1.0 US cents

GLOSSARY

akamatsu	Japanese red pine (*Pinus densiflora* S. et Z.)
akeyama	"open forest;" lord's forest of Owari domain in Kiso that was open to authorized use by locals during Edo period
bubungi	shared-yield tree or trees
buraku	hamlet
buwakebayashi	shared-yield woodland
daimyō	regional barons, lit. "great names," esp. in Edo period
dekasegi	work distant from home, esp. in Edo period
Doitsu tōhi	German spruce (*Picea abies* L. Kasten)
doronoki	a poplar (*Populus maximowiczii* Henry)
ezomatsu	Yezo (Ezo) spruce (*Picea jezoensis* Carr.)
gen'ya	wild grassland, deteriorated woodland
goryō	imperial property
goryōchi	imperial land
Goryōrin	imperial forest
goryō rin'ya	imperial forest and field

hinoki	Japanese cypress (*Chamaecyparis obtusa* Endl.)
hōseirin	Normal Wald (Ger.); a sustained-yield forest
iriai	shared usage
iriaichi	common-use land
iriai dantai	communal association or guild
iriai kumiai	communal association or guild
iriai goryōchi	imperial land with recognized common-use rights
jigyōku	forest operations area
jun ringyōchi	potential timberland, an official forest category
kanrin	centrally-controlled government forest
kan'yūchi	government land
kan'yū rin'ya	government land under prefectural control during Meiji period
kinri goryōchi	imperial property in Edo period
kōgūchi	imperial property in early Meiji
kokuyūrin	national forest
kōyūchi	communal (lit. "public-use") land
kōyūrin	communal woodland
kunugi	an oak (*Quercus acutissima* Carr.)
kurumi	walnut (*Juglandaceae* family; esp. onigurumi or *Juglans* genus)
matsu	pine (*Pinus* sp.)
min'yūchi	non-government land
min'yūrin	non-government woodland
momi	white or silver fir (*Abies firma* S. et Z.)
mura	village
Naichi	in this essay: Honshu, Shikoku, and Kyushu together, as distinct from Hokkaido
nenki	"fixed term," as in fixed-term lease
ohayashi	"lord's forest;" woodland claimed by bakufu or daimyo during Tokugawa period
rinbatsuku	rotation block in forest management program
ringyōchi	timberland, an official forest category
ringyōgaichi	non-timber land, an official forest category
sagemodoshi	"disposal downward" of goryōchi to non-government recipients
sanrin	forest
sanrin gen'ya	woodland and waste
seden goryō	hereditary imperial property
shaji	fanes: Shintō shrines and Buddhist temples
shajirin	fane woodland
shōen	corporate estates, esp. of Heian period
sugi	cryptomeria (*Cryptomeria japonica* D. Don)
todomatsu	white fir (*Abies* sp., esp. *sachalinensis* Fr. Schm. and var. *Mayriana* Miyabe et Kudo)
tsuga	Japanese hemlock (*Tsuga* sp., esp. *sieboldii* Carr.)

yachidamo	an ash (*Fraxinus mandshurica* Rupr. var. *japonica maxim.*)
yamamori	forest warden
zenbatsu	"gradual" or "exploitative selective" logging of designated areas

Appendix B

Forests and the imperial Fisc

Prior to the 1880s, the "restored" Imperial Household was supported by annual allocations from the treasury. During that decade it began to acquire supplemental income-producing properties of its own, and, after 1885, those properties came under the control of the Household Ministry's Property Office (Goryōkyoku).

The general character of the imperial fisc as it was to remain until 1946 was clarified when the Meiji Constitution and Imperial House Law (*Kōshitsu tenpan*) were formulated in 1889–90. The fisc was to consist of three elements: a set annual stipend provided by the government from its regular tax revenues, the yield from a portfolio of securities, and the income from its landed holdings, *goryōchi*.

The imperial stipend was set at 3,000,000 yen in 1890 and raised to 4,500,000 in 1909. The landed estate, as reported in Chapter 2, was mostly assembled in 1889–90. The investment portfolio accumulated gradually, starting in 1885 with stocks in the Bank of Japan and Yokohama Specie Bank, together valued at six million yen. In subsequent years the Household acquired stock in railroads, shipping lines, and other enterprises, and, as the empire grew after 1895, assets in banks and other ventures in Taiwan, Korea, and Manchuria.[1]

The balance in yield among these sources varied from year to year. However, the estimates below, for the years 1916 and 1940, indicate the general proportion of Household income from each source.[2]

Source	1916		1940	
	sum (yen)	%	*sum* (yen)	%
National Treasury	4,500,000	28	4,500,000	16
Goryōchi	3,000,000	19	9,000,000	31
Other Investments	8,500,000	53	15,000,000	53
Total	16,000,000	100	28,500,000	100

Income from imperial landed properties was thus only a part of the whole, but its role grew as decades passed, doing so despite the long-term decline in actual forest acreage.

Even more striking was the growth in imperial assets. Recorded as 517,524 yen in 1875, the sum reached 22,645,583 yen by 1897, almost all acquired during the preceding decade.[3] By 1946 it had risen – in paper value – more than one-hundred fold to some three to six billion yen, a figure that, when adjusted for the wild 1940s inflation, may better be placed at roughly double or perhaps triple the real value of the 1897 figure. Even restated in that manner, the increase was impressive, considering the magnitude of land sales and surrender in the intervening decades and the extent of deforestation and infrastructure loss of the wartime years.

Table 5-1 reports those Imperial Household assets that were held by the Bureau of Imperial Forests at war's end. Those figures compare with the Household's total assets as in the following tabulation:[4]

Asset Category	Bureau of I. F. (1,000 yen)	Household Total (1,000 yen)
Stumpage	1,638,754	1,642,900
Land	635,574	772,620
Buildings	19,586	234,040
Forest Railroads	308,487	308,420
Fine Arts	–	449,790
Vessels (2)	–	10
Securities	–	220,130
Bank Savings, Cash	–	50,150
Debts Receivable	–	32,360
Bonds	–	29,260
Fixtures, Appliances	–	25,620
Miscellaneous	–	10,080
Total	2,602,401	3,775,380

By 1946, then, except for its residual paper assets, valuable villas, palaces, their associated premises, and horde of fine arts, most of the household's material wealth, nearly seventy percent, was associated with its great forest domain.[5]

Appendix C

Goryōrin in Law

In some ways the legal status of Goryōrin was clear; in other ways, it was not. The Imperial House Law (*Kōshitsu tenpan*), which was enacted together with the Meiji Constitution, contained two articles pertaining to Imperial Hereditary Estates (*seden goryō*):

> Article 45. No landed or other property, that has been fixed as the Imperial Hereditary Estates, shall be divided up and alienated.
>
> Article 46. The landed and other property to be included in the Imperial Hereditary Estates, shall be settled by Imperial writ, with the advice of the Privy Council, and shall be announced by the Minister of the Imperial Household.[1]

In fact, of course, *seden goryō*, like non-hereditary property, could be alienated, and much was. In theory, however, that occurred only at imperial behest, and it was handled as a two-step procedure with the parcels in question first reverting to non-hereditary status and being disposed of only after that.

The notion that matters of the Imperial Household properly lay beyond the purview of both Diet and imperial subjects was implicit in the fact that the Imperial House Law was not announced publicly until years after its enactment. As Itō Hirobumi put it, "the Imperial House Law is one that has been settled by the Imperial Family concerning their own affairs," and thus is not a matter of wider concern.[2] And in that spirit, Article 74 of the Law declared that: "No modification of the Imperial House Law shall be required to be submitted to the deliberation of the Imperial Diet."[3] So the House Law was to remain independent of Constitution and Diet.

The imperial fisc, as well, was placed beyond public reach. Article 66 of the Meiji Constitution declared:

> The expenditures of the Imperial House shall be defrayed every year out of the National Treasury, according to the present fixed amount for the same, and shall not require the consent thereto of the Imperial Diet, except in case an increase thereof is found necessary.[4]

Itō went on to explain that the Household need not account for how it used its monies: "Neither consent to these expenditures nor verification of them is required by the Diet." The realm, as noted in Appendix B, provided a set sum (3,000,000 yen in 1890, raised to 4,500,000 in 1909 by Diet action, as per Article 66). The rest of the Imperial Household budget, including income from Goryōrin and investment portfolio, was equally beyond Diet purview, being handled within the Household Ministry.

In terms of litigation, on the other hand, the legal status of the Household and its property proved more ambiguous. In discussing the judicial system of the new Constitutional order, Itō declared that: "The Sovereign is the fountain of justice, and His judicial authority is nothing more than a form of the manifestation of the sovereign power." Further on he stated: "The judicature is combined in the sovereign power of the Emperor as part of His executive power."[5] One could infer, at least, that the Sovereign, being the fount of justice, could not be subjected to its application since doing so would simply place him in judgement of himself.

Elsewhere, in his tortured explanation of the limits of a Japanese subject's "inviolable" property rights, Itō sought to retain what one might call the Sovereign's "right of eminent domain." He explained that the right of a subject's property, "is under the power of the State." To reconcile the subject's right of property with the Sovereign's "supreme right of governing the country," he invoked the wisdom of the celebrated Dutch jurist Hugo Grotius (1583–1645), who held that "a Sovereign possesses the supreme right of property in the land under his rule." And, continued Itō, "Recent writers on the law of nations follow this principle," only replacing the expression "supreme right of property" with the term "territorial sovereignty."[6]

None of this established with certitude whether or not imperial property could properly be subjected to litigation. But it did, at the very least, seem to create an aura of imperial transcendence that could plausibly be extended at least enough to embrace *seden goryō*. And that position, as noted in Chapter 3, was what Household officials chose to affirm, even projecting it to encompass nonhereditary *goryōchi*.

Nor was the Imperial House Law any clearer than the Constitution on the legal status of imperial land. It specified procedures in legal issues involving members of the imperial family but not the Imperial House itself. In consequence, the Household Ministry found the question of whether and how to handle lawsuits regarding Goryōrin a perplexing matter, and while they tried to place it beyond judicial reach, their success in doing so was never fully complete.

Appendix D

Supplemental Figures

FIG. D-1 PREFECTURAL REGULATIONS FOR RENTAL OF GORYŌCHI OR PURCHASE OF STUMPAGE (GUNMA PREFECTURE, 1889)

(Source: Quoted in Tsutsui Michio, *Nihon rinseishi kenkyū josetsu*, pp. 122–3.)

Clause No.1. Although it is stated that no *goryōchi* is to be sold, when the common good requires dikes, embankments, etc. for rivers, roads, or waterways, it may be sold or donated to the extent necessary.

Clause No.2. In cases where someone has rights of land rental, shared-yield forestry, cropping, or other production that predates a site's inclusion in *goryōchi*, those rights shall continue until their term expires. Those who petition for renewal of the term should follow [the following] clauses.

Clause No.3. Those wishing to rent *goryōchi* should proceed as per Prefectural Regulation No.39 of April 1889, which pertains to rental of government land (*kanchi*). However, in that Regulation the articles on non-fee rental in Clauses No.2, 3, 19, and Petition Form No.1 do not apply.

Clause No.4. Those who wish to rent for purposes of tree planting, cattle raising, or harvesting bark, fruit, nuts, or berries should, disregarding the time limits given in Clause No.4 of government land-rental regulations, submit petitions in accord with the stipulations below. Certainly, in regard to tree planting, those clauses on shared yield forestry in the March 1878 Interior Ministry Notice No.4 and those clauses not inappropriate for *goryōchi* in the June 1878 Gunma Prefecture Notice No.14 should be applied.

item: those planting [seedlings for] construction timber, such as pine, *sugi*, or *hinoki*, are to have rental periods of up to thirty years;

item: those planting to produce firewood and charcoal, as with *hinoki* or *kunugi*, are to have up to twenty years;

item: those opening pasture for cattle breeding and raising are to have up to fifteen years;

item: those planting to produce bark or fruit, such as mulberry, paper mulberry, tea, persimmon, peach, or chestnut etc. are to have up to ten years.

Clause No.5. Those who wish to purchase tree or bamboo stumpage should, in a petition based on careful investigation, describe precisely the species of tree desired, the number on the site, and their diameter; attach the proper fee; make marginal note of the land registry (district, town or village, community, neighborhood, acreage, etc.); furnish your birth date certainly; and have two or more guarantors co-sign the petition.

Clause No.6. Those who wish to purchase miscellaneous products or soil or stone should proceed in accord with the Prefecture's Notice No.15 of April 1883 and the application form given in Notice No.3 of January 1889.

FIG. D-2 REGULATIONS FOR FELLING PURCHASED TREES IN GORYŌRIN (BY KITAFUJI ELEVEN-VILLAGE IRIAI KUMIAI, NOVEMBER, 1901)

(Source: Quoted in Hōjō Hiroshi, *Rin'ya iriai no shiteki kenkyū* (jō), pp. 293–294. Herein villagers continue to refer to their villages – which the government called *buraku* – as *mura*. The *mura* of government terminology they identify as "administrative village" (*kanri mura*), as in the first clause.)

1. Anyone in the *iriai kumiai* who wishes to fell purchased trees in Goryōrin must secure a forest-entry license from the administrative-village office and carry it in the forest.
2. The forest-entry license may be used only by the recipient; others cannot borrow it.
3. A license is valid from November 1 to the following October 31.
4. Within that year one may enter the forest [to harvest only] during the following periods:

   January 1 to February last day; June 1 to July 31.
5. When a *kumiai* member with police authority asks to see one's license, one cannot refuse to show it.
6. Those wishing a license must pay fees as follows:

– per license to get out construction timber	3.5 yen
– per license to get out roof boards and rafters	2.5 yen
– per license to get out cartloads of fuel wood	2.0 yen
– per license to get out fuel wood in other ways	1.0 yen

   However, for more than two licenses per household, the fees will be halved.
7. For licenses returned before expiration, there will be no refund.
8. If a licensee loses or damages the license, replacement will cost 10 *sen*.
9. If, in violation of No. 1 or No. 2 above, one enters the forest and harvests without a license or with another's license, the harvest will be confiscated and a fine of one to 1.95 yen imposed.
10. Those who violate the time limits in No. 4 will be denied forest entry for one year.
11. Those who refuse to show a license as per No. 5 will be fined from 50 *sen* to 1.95 yen.

12. Those who in violation of No. 8 neglect to get a new license or to carry it will be fined 5 to 50 *sen*.
13. Clauses No. 9 to No. 12 will be handled by the heads of *kumiai* villages, and *kumiai* villages will receive any confiscated goods and fines.
14. To protect the forest, each village will name a forest warden (*yamamori*) and patrol against illegal felling and theft.
15. Anyone who detects a forest fire should warn the village at once, and all villagers must turn out to quell the blaze.

Appendix E

Supplemental Tables

Table E-1: Total Woodland of Japan, 1939

Region	*Category*	*Acreage (1000 chōbu)*	*Stumpage (million koku)*			*Koku per chōbu*
			Needle-leaf	*Broad-leaf*	*Total*	
Naichi	Kokuyūrin	4,203	524	1,167	1,691	402
	Goryōrin	486	124	50	174	358
	Min'yūrin	12,886	1,731	970	2,701	210
	Total	17,575	2,379	2,187	4,566	260
Hokkaido	Kokuyūrin	3,510	580	749	1,329	379
	Goryōrin	894	116	271	388	434
	Min'yūrin	2,207	95	312	407	184
	Total	6,611	791	1,332	2,124	321
Grand Totals	Kokuyūrin	7,713	1,104	1,916	3,020	392
	Goryōrin	1,380	240	321	562	407
	Min'yūrin	15,093	1,825	1,282	3,107	206
	Total	24,186	3,169	3,519	6,689	277

(Source: Hagino Toshio, *Nihon gendai rinsei no sengō katei*, p. 42. The column "*Koku* per *chōbu*" is added here. I have deleted three digits in all of Hagino's columns. Also I have recalculated the Grand Totals for *kokuyūrin* because Hagino included in them Karafuto, his figures for Karafuto being 2,928, 654, 93, and 747, respectively. Forestry works published before the Pacific War commonly included the colonial possessions of Karafuto, Taiwan, and Korea. Most of those published since 1945 have omitted them. Hagino included only Karafuto, it appears, because its woodland was administered directly as *kokuyūrin*, whereas that in Korea and Taiwan was overseen by the respective colonial government.)

Table E-2: Location of Goryōrin, 1937
(listed in descending order by *chōbu* per prefecture; with select 1889 comparisons)

Prefecture	Hereditary	Regular	Total*	Select 1889
Hokkaido		906,604		
Nagano	112,535	49,812	162,347	144,927
Shizuoka	39,969	62,533	102,502	210,180
Gifu	45,003	12,980	57,983	158,395
Iwate		41,550		
Aomori		31,888		
Gunma		30,661		
Tochigi		20,683		
Yamagata		18,101		
Aichi	6,090	10,611	16,701	76,626
Kanagawa	4,881	2,731	7,612	25,276
Mie		6,115		
Chiba		6,042		
Fukushima		2,491		
Tokyo		1,780		
Niigata		181		
Ibaraki		61		
Osaka		1		
Yamanashi			0	412,579
Total	208,478	1,204,825	1,413,303	

* Given where Total exceeds Regular.

(Sources: 1937 figures based on Tsutsui Michio, *Nihon rinseishi kenkyū josetsu*, p. 64. Selected 1889 figures from Hōjō Hiroshi, *Rin'ya iriai no shiteki kenkyū (jō)*, p. 266.)

Table E-3: Property Office Regional Headquarters, 1914
(location and supervised acreage)

Branch	*Registered Acreage (chōbu)*	*Surveyed Acreage (chōbu)*
Sapporo	651,673	943,813
Tokyo	358,128	246,606
Nagoya	401,674	159,305
Kiso	470,450	170,186
Total	1,881,925	1,519,910

(Source: Matsunami Hidemi, *Meiji ringyō shiyō kōshū*, Vol. II, p. 1033. Matsunami also included Kyoto, where an office oversaw 629 *chōbu* of *goryōchi* (of which 456 was surveyed), evidently palace grounds and tomb sites.)

Table E-4: Rotation-Cutting Schedule of Hundred Year Plan

Procedure	*Total Area* (*chōbu*)	*Honshu*			*Hokkaido*		
		Species	Cycle (years)	Acreage (*chōbu*)	Species	Cycle (years)	Acreage (*chōbu*)
		sugi/hinoki	80–90	110,000	ezomatsu/todomatsu	80	100,000
Clear-cutting	302,000	matsu	60	50,000			
		sugi	120	42,000			
		Total		*202,000*			*100,000*
=========							
		sugi	120	44,000	ezomatsu/todomatsu	100	456,000
		shrine timber		8,000	broadleaf deciduous	120	240,000
Selective Cutting	798,000	momi/tsuga	120	50,000			
		Total		*102,000*			*696,000*

(Source: Derived from Tsutsui Michio, *Nihon rinseishi kenkyū josetsu*, p. 144. Tsutsui lists *"kanematsu"* (or *kinmatsu, kanamatsu*) with 120-year *sugi*, but I was unable to identify it.)

Table E-5: Goryōrin Management Results 1908–38 (in hectares)

	1908	1918	1928	1938
Total Managed Area	354,337	744,657	1,221,384	1,301,030
Total Ringyōchi Area	296,632	533,352	942,884	1,090,889
As % of managed	84%	72%	77%	84%
Hokkaido Ringyōchi	37,503	232,987	657,006	784,466
Total Clearcut Area	179,510	249,504	252,789	260,177
Total Selective Cut	70,195	235,595	630,582	727,764

(Source: Derived from Ringyō hattatsushi chōsakai, comp., *Nihon ringyō hattatsushi (jō)*, p. 462, Tables 37, 38.)

Table E-6: Trends in Goryōrin Logging Practice, 1908–38

	1908	1918	1928	1938
Tokyo				
Clearcut (%)	32.3	90.6	93.2	79.6
Selective (%)	6.9	3.7	3.0	20.0
Total (ha)	42,556	97,593	97,268	96,496
Nagoya				
Clearcut (%)	94.1	84.8	76.7	61.0
Selective (%)	0.5	12.3	23.2	48.9
Total (ha)	86,304	108,596	99,528	107,306
Kiso				
Clearcut (%)	69.5	80.7	85.0	51.8
Selective (%)	24.0	19.2	14.9	43.6
Total (ha)	121,342	85,334	80,043	92,671
Hokkaido				
Clearcut (%)	0	0	2.6	9.1
Selective (%)	100.0	86.7	90.1	81.7
Total (ha)	37,504	232,987	657,005	765,638

(Source: Ringyō hattatsushi chōsakai, comp., *Nihon ringyō hattatsushi (jō)*, p. 463. The original Table 39 gives both acreage and percent figures for all categories. And it lists, in addition to clearcut and selective logging, three other minor categories of operations to yield 100% figures in the TOTALS. Of these three, "gradualist felling" (*zenbatsu*) accounted for 59% in the Tokyo regional headquarters' region in 1908 and stood in the 7–14% range in Hokkaido from 1918 onward, while "brushwood" (*wairin*) was a small but persistent portion of Naichi acreage.)

Table E-7: Trends in Goryōrin Stands, 1909–38 (in cubic meters)*
* evidently m^3 per hectare for stumpage. Method of estimate unclear for annual increment and harvest volume.

	Kiso	Nagoya	Tokyo	Sapporo	Average
Stumpage Volume					
1909	264.5	164.6	95.7	155.7	197.3
1918	267.9	190.3	53.0	145.9	159.3
1928	270.6	185.1	52.0	142.0	149.3
1938	264.5	174.6	67.6	140.7	146.0
Annual Increment					
1909	1.92	1.72	1.24	1.55	1.71
1918	1.62	1.80	0.77	1.29	1.36
1928	1.69	1.46	0.93	1.00	1.20
1938	1.60	1.54	1.54	1.14	1.23
Harvest Volume					
1909	3.16	2.39	2.53	1.75	2.66
1918	3.33	2.57	1.38	1.68	2.10
1928	2.90	2.43	1.36	1.47	1.69
1938	3.23	2.47	1.56	1.58	1.78

(Source: Ringyō hattatsushi chōsakai, comp., *Nihon ringyō hattatsushi (jō)*, p. 481. (Derived from *Teishitsu rin'yakyoku gojūnenshi*, p. 653, without explanation) The original Table 43 also lists Asahikawa, a regional office established in Hokkaido after 1928, giving its figures for 1938 (132.6 m^3 of stumpage; 1.06 of growth, and 1.47 of harvest). Their omission here may alter the 1938 averages slightly.)

Table E-8: Goryōrin Afforestation, 1906–15 (acreage and seedlings)

Year	Hinoki		Total	
	Acreage (*chōbu*)	*Seedlings* (in 1000s)	*Acreage* (*chōbu*)	*Seedlings* (in 1000s)
1906	773	4,203	2,848	14,799
1907	1,162	4,344	3,126	15,028
1908	970	5,513	2,789	16,170
1909	1,073	5,692	2,761	15,590
1910	786	4,407	2,180	12,960
1911	929	5,141	2,644	15,575
1912	1,053	5,710	2,629	15,536
1913	1,193	6,459	3,294	18,733
1914	1,888	9,097	4,488	23,353
1915	2,454	12,216	4,510	23,943

(Source: Matsunami Hidemi, *Meiji ringyō shiyō kōshū*, Vol. II, Appendices, pp. 24–5. Pine and cryptomeria were the two main other species that Matsunami identifies, but he also lists six other species plus the largest category of all, "other.")

Table E-9: Goryōrin Afforestation, 1936–45 (acreage)

Year	Planted (*chōbu*)	Natural (*chōbu*)	Total (*chōbu*)
1936	5,229	8,144	13,372
1937	4,286	12,472	16,758
1938	5,056	13,606	18,674
1939	4,420	20,603	25,023
1940	4,613	24,833	29,446
1941	3,438	24,917	28,355
1942	2,908	22,157	25,065
1943	2,982	20,072	23,054
1944	2,897	11,161	14,058
1945	1,606	5,522	7,128

(Source: Kuroda Hisata, *Tennōke no zaisan*, Table 5–2, p. 251 (addition as in original).)

Table E-10: Trends in Goryōrin Logging, 1892–1937 (seven-year annual averages)

Period	Average Annual *Cut (m³)*	Logging *Strategy (%)*		Uses of *Yield (%)*	
		Government Felling	Stumpage Sale	Lumber	Fuel
1892–98	453,805	9.5	90.4	52.3	47.6
99–1905	728,763	6.8	93.1	59.8	40.1
1906–12	1,165,687	14.6	85.3	67.5	32.4
1913–19	1,321,801	16.2	83.7	68.3	31.6
1920–26	1,473,103	21.7	78.2	70.2	29.7
1927–31	1,670,241	21.7	78.2	70.8	29.1
1932–37	1,922,116	30.1	69.8	70.3	29.6

(Source: Ringyō hattatsushi chōsakai, comp., *Nihon ringyō hattatsushi (jō)*, p. 477.)

Table E-11: Goryōrin Harvest, 1936–45

Year	Government Logging				Stumpage Sales .	
	Logs (*koku*)	Lumber (*koku*)	Charcoal (1,000 *kan*)	Firewood (*koku*)	Saw timber (*koku*)	Fuelwood (*koku*)
1936	1,515,344	–	1,645	63,647	2,494,098	2,021,058
1937	1,639,265	–	1,697	65,861	2,757,442	1,852,456
1938	1,889,165	–	1,632	–	2,561,515	1,604,790
1939	2,230,032	–	1,732	–	2,618,773	1,711,195
1940	2,710,008	–	3,357	–	2,384,481	1,542,424
1941	2,469,192	7,461	4,485	92,292	2,534,818	1,591,801
1942	3,009,819	9,798	4,582	132,899	2,388,283	1,459,368
1943	4,237,393	7,260	5,080	127,987	1,990,400	1,208,012
1944	4,910,974	20,590	5,879	188,184	4,855,633	996,700
1945	4,858,931	25,450	3,906	87,443	1,676,344	989,240

(Source: Hagino Toshio, *Nihon gendai rinsei no sengō katei*, Table 8, p. 62.)

Table E-12: Goryōrin Stumpage, 1939 and 1946

Place	Year	*Acreage* (*chōbu*)	*Needleleaf* (1000 *koku*)	*Broadleaf* (1000 *koku*)	*Total* (1000 *koku*)
Naichi	1939	486,402	124,013	49,673	173,686
	1946	428,121	176,806	58,240	235,046
Hokkaido	1939	893,685	116,375	271,458	387,833
	1946	878,006	124,899	285,138	410,037
Total	1939	1,380,087	240,388	321,131	551,519
	1946	1,306,127	301,705	343,378	645,083

(Source: Taken from Hagino Toshio, *Nihon gendai rinsei no sengō katei*, p. 42, Table 2; p. 59, Table 4.)

Table E-13: Goryōrin Finances, 1889–1947 (Five-year averages; in 1000 yen)

Period	Income	Expenditure	Balance	Profit Rate	Land Sales Income	Total
1889–93	459	661	−202	–	1	−200
1894–98	813	578	234	28.8%	386	621
99–1903	1,206	781	425	35.2	273	698
1904–08	1,574	1,002	572	36.4	393	965
1909–13	2,942	1,657	1,285	43.7	663	1,948
1914–18	5,794	2,350	3,444	59.4	368	3,812
1919–23	14,058	6,300	7,758	55.2	786	8,544
1924–28	14,079	7,335	6,744	47.9	3,023	9,767
1929–33	9,764	6,186	3,578	36.6	1,425	5,003
1934–38	14,722	8,581	6,141	41.7	1,569	7,710
1939–43	53,999	30,051	23,948	44.3		
1944–47	158,049	123,337	34,712	22.0		

(Sources: Based on unrounded figures in Ringyō hattatsushi chōsakai, comp., *Nihon ringyō hattatsushi (jō)*, p. 484, for 1889–1938; calculated from Hagino Toshio, *Nihon gendai rinsei no sengō katei*, p. 60, Table 6, for 1939-March 1947. "Profit Rate" (balance as percent of income) figures added here.)

Table E-14: Government Forest Profit Margins, 1936–45 (annual average; in 1000 yen)

Category	Income	Expenditure	Profit		
			(net profit)	(as % of total net)	(as % of income)
Naichi kokuyūrin	113,105	100,437	12,668	21.9	11.2
Hokkaido kokuyūrin	45,435	21,450	23,986	41.5	52.8
Subtotal	158,540	121,887	36,654	63.4	23.1
Goryōrin	52,658	31,527	21,131	36.6	40.1
Total	211,198	153,414	57,785	100.0	27.4

(Source: Hagino Toshio, *Nihon gendai rinsei no sengō katei*, p. 59, Table 5. Hagino gives percent of total net profit. Subtotal for *kokuyūrin* and profit as percent of income are added here.)

Notes

NOTE: In these endnotes, references are cited in the social science manner: author's surname and date of publication. For full identification, consult bibliography.

Part I

Preface

[1] Use of the Anglicism "Imperial" rather than "Royal" in referring to Japan's royal house conforms to standard usage. It reflects not only the political history of that monarchy during the past 150 years when it struggled to establish itself among the "Great Powers," but also its original quest some 1,500 years ago for a proper title vis-à-vis the Emperor of China. At that time government leaders employed the unusual Chinese term *tennō,* "heavenly sovereign," to denote their monarch. That title avoided terminological conflict with China's emperor while sidestepping the subordinate status that other more regular Chinese kingly titles entailed.

[2] Japan encompasses some 145,000 square miles; the UK and Ireland combined, some 120,000.

[3] The material on Japan in this Preface is treated somewhat more fully in Totman, 2005. On Britain's woodland see the wonderfully rich study of Scottish forest history, Anderson, 1967; the solidly detailed treatment of English forest history, James, 1981; and three shorter but useful and readable works: Havins, 1976, on England; Linnard, 1982, on Wales; and McCracken, 1971, on Ireland. Sampson, 1965, limns the role of the twentieth-century monarchy.

[4] For another, although less directly relevant, similarity, it may bear note that certain ethnic groups in or adjacent to both Britain and Japan ultimately proved stronger than others and in due course overran or momentarily conquered them – Scots, Welsh, and Irish in the English case; Ainu, Ryukyuans and Koreans in the Japanese. Today, in consequence, both societies face unresolved ethnic conflicts and both exemplify the difficulty such conflicts present to nearly all the world's peoples, be they in the Balkans, Near East, Indonesia, the Americas, or wherever. More to the point here, in both cases, the overrunning

of neighboring societies was undertaken by, or in the name of, monarchy. In consequence the issues of ethnic conflict and monarchial legitimacy are intertwined, which has a bearing on public and scholarly attitudes toward the monarchies and their roles.

5 These comments on plate tectonics follow Bambach, 1980, pp. 86–98, and Dietz, 1970, pp. 30–41.

6 The Geological Survey of Japan has mapped these plate boundaries. See the sketch map reproduced in Totman, 2005, p. 564. Chapter 1 of Totman, 2004, treats this geological history a bit more fully.

7 A handy reference on demographic history is McEvedy, 1978.

8 The phrase "to one or another degree" is important. Japan, like the British Isles, has over the centuries suffered from much forest deterioration. Today much of that 70% is indeed well forested terrain, but in the mid-nineteenth century perhaps half of Naichi woodland had few if any trees of appreciable size. And as the following chapters suggest, forest recovery was a slow and difficult task prior to about 1960.

9 James, 1981, p. 257. See *The Ordnance Survey*, 1982, for woodland locations.

10 On forest biodiversity, see Heske, 1938, p. 52, and Simmons, 1979, p. 114.

11 Concurrently, as we shall note en passant, there was developing in Japan, much as in Britain, a concern for timber production and riparian management that transcended ownership categories. This development of the 1870s onward appears most clearly, however, when one examines the activities of Japanese government and entrepreneurial foresters and their professional organizations, so it is a story for another time.

Introduction

1 The term *goryōrin* should, of course, be italicized. But in this essay I Anglicize it as a proper noun because it appears in the title and is so prominent in the text.

2 Initially called Goryōkyoku, what I call the Property Office was retitled Teishitsu rin'ya kanrikyoku (Supervisory Office of the Imperial Forest) in 1907, when the administering of imperial woodland was separated from the handling of other imperial properties, such as palaces, and the Imperial Household's overall fisc. Later, in 1924, the title was simplified to Teishitsu rin'yakyoku – Bureau of the Imperial Forest. For the sake of verbal consistency, however, this essay uses the term Property Office throughout.

3 One suspects, too, that the endemic center-periphery conflict over control and use of woodland was a significant propellant in the later Meiji rise of "hometown pride" that Wigen, 2000, explores so nicely in the case of Nagano Prefecture.

4 The most solid English-language study of the prewar imperial institution is still the pioneering work, Titus, 1974. Two recent and valuable scholarly biographies that study the emperor's political role are Large, 1992, and Wetzler, 1998, both of which contain good bibliographical guidance to other works. Two large works in the *J'accuse* tradition that focus on the hoary agenda of imperial culpability for the Pacific War are Bergamini, 1971, and Bix, 2000. A gossipy popular treatment in the manner of journalistic approaches to the British royal family is Seagrave, 1999.

5 Regarding the term *kokuyūrin*, a semantic clarification seems in order. Meiji-era writers used various terms – *kanrin, kan'yūrin, kan'yū rin'ya, kan'yū sanrin gen'ya, kokuyūrin, kokuyū rin'ya*, etc. – to denote government woodland. For most of the Meiji period, however, a substantial part of that forest was administered by prefectural officials while a growing portion was handled by the central Forest Agency. Scholars commonly use the term *kanrin* to denote centrally-controlled government woodland and *kan'yūrin* (or *kan'yū rin'ya*) to denote the other. *Kokuyūrin*, which became the official term for government forest in 1897, serves in this essay to denote *kanrin* and *kan'yūrin* combined.

6 On Japan's forests since World War II a valuable recent work is Yoshiya, 2002.

Chapter 1 Historical Background

1 Tsunoda, 1958, pp. 106–107. Although "crown lands" (*miyake*) had existed before the seventh century, those modest imperial holdings were not what early Meiji leaders had in mind as precedent for their monarch.

2 Hall,1966, p. 343. For more detail, see Webb, 1968, pp. 126–8. A recent study of the Imperial Court during these centuries of difficulty is Butler, 2002. Butler seeks to highlight the revival of courtly influence after the 1570s but doesn't address the issue of land control.

3 See Titus, 1974, pp. 13–49 on this thorny matter.

4 On matters of ritual, see the delightfully illustrated treatment in Fujitani, 1998.

5 Hōjō, 1977, p. 85. Kuroda, 1966, pp. 18–19, 37–8, 50, 83–4, identifies palace-related sites. The terms *chōbu* and *chō* are synonyms, but the same *chō* is also a linear measure (109 m), so I use *chōbu*, rather than *chō*, to denote a "square *chō*." A warning: one must treat the numbers in this essay cautiously. The acreage figures in particular are unreliable and inconsistent, with different sources, both primary and secondary, giving different figures. The reasons are several: (1) many figures, especially from early and mid-Meiji, were guesses; others, the result of imprecision or error in mensuration; (2) resurveys continually produced new figures; (3) different institutions compiled and consolidated figures differently; (4) changes in usage and modifications in forest categories produced statistical changes; (5) some sources round off figures; others do not; (6) scholarly error.

6 Part II (Chapter 8) treats this issue in detail.

7 Kozo Yamamura, "The Meiji Land Tax Reform and Its Effects," in Jansen, 1986, p. 385, sums up nicely the 1869 proposals of the "Dutch Scholar," Kanda Takahira (Kōhei), which foreshadowed the policy on property ownership implemented in following years. The issue of use rights and property rights is a prickly one among legal scholars. However, much of the prickliness stems from the notion that the latter is a more "modern" or "advanced" legal concept than the former, which has prompted some scholars to seek to demonstrate its existence in Japan before the arrival of European influence. See works of Hōjō Hiroshi, in particular, which treat the topic at length and which also will guide one to other scholarship. In some recent scholarship, however, this notion of "modern" (*kindaiteki*) absolute ownership rights is treated as a distinctively Meiji-era concept that began to be replaced in the 1920s by a more progressive "recent" (*gendaiteki*) ideal of socially

responsible ownership, which, it is argued, flowered after 1945 and is now leading to a more ecologically-conscious approach to property. For this perspective, see works of Hagino Toshio but also Mitsui Shōji, notably his essay, Mitsui, 1994.

8 On tax reform and the technical task of determining the value of land, see Nakamura, 1966, Appendix A, pp. 177–96, and Yamamura in Jansen, 1986, pp 382–9.

9 The term "forest and fen," its lovely conciseness notwithstanding, does not quite fit *sanrin gen'ya* because Japan was spared the glacial-age land-surface disruptions that created much of the fen – swamps and marshland – of the British Isles.

Chapter 2 Establishing Goryōrin

1 For example, Beasley, 1963, Chapters 6–8, provides concise and straightforward coverage of this history.

2 On these factional matters, see Lebra, 1973.

3 Ōkuma's long memorial is translated in Centre, 1972, Vol. 3, pp. 46–60. Another translation appears in Beckmann, 1957, pp. 136–42. Lebra, 1973, pp. 41–2, adumbrates its main points.

4 Translated in Centre, 1972, Vol. 3, pp. 69–70. Also in Beckmann, 1957, p. 149.

5 Besides Beckmann, 1957, two other careful studies of this constitution-making process are Akita, 1967, and Banno, 1992. Also useful for its close examination of legal niceties is Siemes, 1968.

6 On this topic see, in particular, Bowen, 1980.

7 The following material on Matsuno in Europe comes from Nishio, 1988, pp. 41–3. Standard references give the name as Matsuno Kan; Hazama is given in *furigana* at p. 372 of Nihon ringyō, 1997. Besides having a distinguished career in forestry, Matsuno was known for his pioneering role in inter-cultural marriage. He wed the brave and venturesome Clara Tietelmann, whom he met in Germany and who went on to her own career as founder of a kindergarten in Tokyo. See Kido, 1983, Vol. 3, pp. 348, 350–2, 406–407. In his diary entries during his weeks in Berlin (March 7 – April 26, 1873, save for fifteen days in Russia: Vol. II, pp. 296–318), Kido does not mention Matsuno by name. But he records meeting Aoki many times, meeting groups of Japanese students on a couple of occasions, and meeting Prince Kitashirakawa, whom Matsuno accompanied to Europe with the original intent of studying general economics, three times.

8 James, 1981, pp. 194–6, adumbrates this matter.

9 Surely, too, the influence of German thought in forestry was enhanced after 1878 by the role of the German scholar, Hermann Roesler, in the development of the Meiji Constitution and government policy. On Roesler's role, see Siemes, 1968.

10 Narita, 1996, p. 61.

11 For a concise discussion of this European influence, see Tsutsui, 1987, pp. 72–5.

12 Ringyō hattatsushi, 1960, pp. 202–205 summarizes the formation of *sanrinkai*.

13 Lebra, 1973, p. 33, adumbrates its main points.

14 Ōkuma is quoted in Narita, 1996, p. 61.

15 Ringyō hattatsushi, 1960, pp. 100–101. The Nōshōmushō official was First Secretary Wakayama Giichi. Kuroda, 1966, pp, 23–4, quotes Iwakura at length.

16 Itō saw *goryōchi* as a political tool as well as a fiscal one. In a memorandum of 1885, he conceded the centrality of Hokkaido with its "fertile soil" and Kiso with its "fine timber." But he also argued for assigning to the Imperial Household the many historical sites associated with its glorious history, to preserve them and make them available to visitors. Doing so, he averred, would help keep members of the imperial family aware of their legacy and purpose, would heighten public loyalty and devotion to the imperial realm, and would enable foreign visitors to learn and appreciate the imperial legacy. One suspects that Itō's memo reflected his own visits to historic sites in Europe. Quoted in Hōjō, 1983, pp. 465–7.

17 See Hōjō, 1983, pp. 463–75, for extensive quotation of opinions favoring creation of Goryōrin during the later 1880s.

18 For the figures, see Kuroda, 1966, pp. 247–8, Table 2, which lists the annual financial figures represented here by Figure 5–5 in Chapter 5.

19 This multi-asset *goryōchi* is examined in Ringyō hattatsushi, 1960, pp. 132–3, 425–37. On its use as arable and pasturage, see Matsunami, 1919, Vol. 1, pp. 240–2, 260–1.

20 The survey instructions of November 24, 1892 are given in Matsunami, 1919, Vol. 2, pp. 1043–5.

21 Matsunami, 1919, Vol. 2, pp. 1034–5.

22 This summary of the creation of Goryōrin derives mainly from Ringyō hattatsushi, 1960, pp. 103–113, which has good tables listing the parcels acquired. That work in turn relies heavily on Teishitsu rin'yakyoku, 1939.

23 By one calculation, in 1893, after the Household Ministry released much of its Hokkaido claim, Goryōrin was registered at 2,261,000 *chōbu, kokuyūrin* at 19,179,000, and *min'yū rin'ya* at 8,414,000. Hōjō, 1977, p. 102, Table 2. The total (29,854,000 *chōbu*) is excessive (see Table Int-1 in the Introduction for later, more reliable figures), and most of the overestimate was in *kokuyūrin,* which included most of the poorly measured interior mountain regions of Naichi and Hokkaido.

24 The passage is taken from Hōjō, 1977, p. 91, quoting Teishitsu rin'yakyoku, 1939.

25 "Naichi" is a relativistic term whose meaning depends upon its context. Authors discussing the empire, which included Taiwan, Korea, and Karafuto, often place Hokkaido within Naichi. But when authors wish to distinguish Hokkaido from the older realm of Honshu-Shikoku-Kyushu, they commonly use Naichi to denote the latter. At times they seek to reduce the terminological ambiguity by using the awkward construction *naichi fuken* to denote the latter, juxtaposing the *fu* and *ken* prefectural units of Honshu, Shikoku, and Kyushu against the unique *dōchō* or "circuit" administrative unit of Hokkaido.

26 Itō, 1906, p. 180. The translation also appears in Centre, 1972, Vol. 3, p. 111.

Chapter 3 Reducing Goryōrin

1 These figures come from Kuroda, 1966, p. 109. He gives other statistics, together with *goryōchi* place names, on pp. 37–8, 46–50, 83–6. Other reasonably similar figures appear in Ringyō hattatsushi, 1960, and Hagino, 1996. The 1937 figure is from Tsutsui, 1978, p. 64. Matsunami, 1919, Vol 1, pp. 122–4, offers a concise summary of the losses.

2 Tsutsui, 1978, p. 64.

3 Hagino, 1990, pp. 333–5.

4 Hagino, 1990, p. 31, mentions the Shōwa emperor's query.

5 Hagino, 1996, pp. 56–7. Hagino, 1993, explores wartime forest matters in detail but does not treat Goryōrin as such.

6 See Hōjō, 1983, pp. 603–77, for the documents of this case. On later developments in Kiso Goryōrin, and a guide to source materials, see Shimizu, 1986. I note briefly the earlier stages of the "Kiso incident" in Chapter 8.

7 Tsutsui, 1978, pp. 126–7.

8 Ōgane, 1973, pp. 55–7.

9 Dai Nihon, 1983, pp. 79, 80, 82.

10 Kuroda, 1966, pp. 57, 90–1, 111, 113.

11 A concise treatment of the 1899–1901 legislation and Household Ministry policy is Ringyō hattatsushi, 1958, pp. 129–31. A fuller treatment of the legislative story is Hagino, 1990, pp. 331–48, which mentions the Household attitude en passant.

12 Bix, 2000, employs the customary dyadic public/private analytical categories, characterizing Goryōrin as private land. That construction fits his purposes but conceals the legal ambiguity that the category of Goryōrin presented to authorities.

13 Tsutsui, 1978, pp. 130–1.

14 Kuroda, 1966, pp. 248–9, Table 2-1, lists annual land-income figures.

15 This summary is based on Ringyō hattatsushi, 1958, pp. 439–56.

16 This acreage appears to include that of the Ise Shrine and Mt. Takachiho, mentioned in Chapter 2.

17 Bix, 2000, pp. 94–5, reports that in October 1919 Yamagata Aritomo urged the Imperial Household Ministry to dispose of wetlands and dry fields. It was a safe admonition, it appears, because in July 1918 a Ministry committee had decided to dispose of its agricultural landholdings, which had proven unprofitable, and by year's end the Ministry had issued the regulations for their sale, which commenced during 1919. Ringyō hattatsushi, 1960, pp. 449–51.

Chapter 4 The case of Yamanashi

1 Smethurst, 1986, p. 106.

2 Before the 1870s, the integrative effect of Yamanashi's central location between the urban hubs of Tokyo and Nagoya was more than offset by the isolationist effect of its severe topography. Roughly speaking, the frontiers of later-Tokugawa regenerative forestry and entrepreneurial lumbering pressed westward from Edo nearly to the eastern boundary of Kai Province (later Yamanashi Prefecture), northward from Suruga Bay into the southern arm of Kai southwest of Mt. Fuji, and northward up the Tenryū River into the Ina

district of Shinano Province (later Nagano) on the western flank of Kai. So Kai proper mostly lay beyond their reach.

3 Tsutsui, 1978, pp. 92–4, treats 1880s Yamanashi woodland concisely. Rin'ya seido, 1977, contains numerous documents relating to this Yamanashi forest history.

4 The richest study in English of prewar Yamanashi and its sericulture is Smethurst, 1986. His study focuses on six villages in the agriculturally most productive part of the Kōfu Basin. For a concise look at the evolution of agriforestry in far southern Yamanashi during these decades, see Ringyō hattatsushi, 1958, pp. 555–8. A fine study of contemporary economic change in the adjacent Ina Valley of Nagano Prefecture is Wigen, 2000.

5 Tsutsui, 1978, p. 93. Rin'ya seido, 1977, pp. 221–2, quotes this document and the Nōshōmushō reply. But in what appears to be a typographical error, it gives the governor's proposal for the shorter lease as "ten years" rather than twenty. However, the Nōshōmushō reply "approves" it as twenty, with no indication that their number was a revision, whereas it explicitly notes the change of the longer lease from fifty to twenty.

6 Rin'ya seido, 1977, pp. 274–7.

7 Hōjō, 1968, p. 41, gives these figures, which total 414,959. Hōjō, 1977, pp. 102, 265, and Ringyō hattatsushi, 1958, p. 105, give a total of 412,579. But on p. 121, Ringyō hattatsushi, 1958, uses the figure 406,451, nicely revealing the dicey nature of the numbers.

8 Documents in Rin'ya seido, 1977, pp. 489–96, illustrate this relationship.

9 Quoted in Ringyō hattatsushi, 1960, p. 121.

10 Hōjō, 1968, pp. 42–4, discusses those aspects of these regulations that relate to local use rights.

11 Tsutsui, 1978, pp. 95–7, presents the eighteen-clause regulations. Hōjō, 1977, pp. 267–8, gives a sixteen-clause version that differs slightly in some particulars.

12 Ringyō hattatsushi, 1958, pp. 157–65, outlines this process.

13 Rin'ya seido, 1977, pp. 501–505, reproduces the regulations.

14 See Hōjō, 1977, pp. 270–5, 286–9, for examples of village *iriai* agreements.

15 Ōshima,1976, p. 125.

16 Ōshima, 1976, pp. 143–7.

17 Ericson, 1996, pp. 30, 46.

18 Tsutsui, 1978, p. 99.

19 Tsutsui, 1978, p. 100.

20 Rin'ya seido, 1977, pp. 511–15.

21 Tsutsui, 1978, pp. 102–104.

22 Hōjō, 1977, pp. 282–5.

23 After World War II the American army inherited and used this military training area (the Kitafuji area, on the northern slope of Mt. Fuji), and an incident in which local residents were killed and injured by exploding ordnance while scavenging spent shell casings led to court cases in which the question of rights of access was reopened. Those 1950s cases spurred the assemblage of documentation that facilitated the later scholarly study of Kitafuji *iriaichi*.

24 Hōjō, 1977, pp. 295–301.

25 In some prefectures, substantial extra logging was done in conjunction with the Russo-Japanese War, and it seems likely that such occurred in Yamanashi as well, adding to forest vulnerability.

26 Ringyō hattatsushi, 1960, pp. 123–4. Regulations also quoted in Tsutsui, 1978, p. 98.

27 Tsutsui, 1978, pp. 104–105.

28 Ringyō hattatsushi, 1960, p. 124.

29 Smethurst, *Agricultural Development and Tenancy Disputes,* p. 109.

30 Ringyō hattatsushi, 1960, p. 734.

31 Ringyō hattatsushi, 1960, pp. 734–6.

32 Kuroda, 1966, reports that in the 1930s there were 18,000 (p. 48) or 10,000 (p. 111) *chōbu* of Goryōrin in Yamanashi. Whether this was a surviving portion, or a later restoration is unclear.

Chapter 5 Managing Goryōrin: Plans and Performance

1 Matsunami, 1919, Vol. 2, pp. 1032–3, summarizes organizational changes to 1914.

2 Note, however, that some sources include the Kyoto office, which handled palace, villa, and tomb sites, as a regional headquarters, and that between 1928 and 1938 another office was opened at Asahikawa in Hokkaido. See notes to Tables E-3 and E-7.

3 Hagino, 1996, p. 61. Hagino does not identify the six regional headquarters. Titus, 1974, pp. 2, 61.

4 Hagino, 1996, pp. 66–7.

5 This section on management plans is mostly summarized from Ringyō hattatsushi, 1960, pp. 460–2. But on the Hundred Year Plan, see Tsutsui, 1978, pp. 143–4.

6 Tsutsui, 1978, p. 144. The plan included the final cut of rotation cutting in its selective cutting (*takubatsu*) category, perhaps because natural rejuvenation on such sites would already be in progress due to natural reseeding in patches opened up by earlier thinning work.

7 Yield tables provide estimates of the timber volume on a site, based on standardized criteria (location, tree species, age, trunk diameter at breast height, stem height, and stand density). There are several methods for making estimates, as explained, for example, in Daniel, 1979, pp. 259ff. An example of Japanese yield tables is Tokyo kyōiku, 1955.

8 This is a paraphrase of wording in Teishitsu rin'yakyoku, 1939, as quoted in Dai Nihon, 1983, p. 559.

9 Matsunami, 1919, Vol. I, p. 121.

10 Ringyō hattatsushi, 1960, pp. 457, 460.

11 Figures from Dai Nihon, 1983, p. 563, Table 43.

12 Hagino, 1993, p. 29. What portion, if any, of this Goryōrin logging was performed by captive labor is unclear, but in the next two years such labor became a more pronounced part of the general work force (particularly in mining, where it was easier to police).

13 The following material on Sanru comes from Ōgane, 1973, pp. 60–2, 68–70.

14 The formula used in arriving at these figures is unclear to me. But it probably was similar to the relatively simple equation that the Property Office had employed in harvest guidelines of 1897. That equation was E = 2M/n+1, in which E was the volume of harvest; M, the stand accumulation, and n, the stand rotation (e.g. 120 years) divided by thinning cut cycle (e.g. forty years), plus one. Thus on a site with a stumpage estimate of ten million *koku,* and with thinning and final cuts to be made in forty-year cycles over a 120-year stand life, the formula would give E = 2 x 10,000,000/(120/40)+1, which equals twenty million/3+1, or five million *koku* per cut. Ōgane, 1973, p. 54, discusses this matter briefly.

15 Ōgane, 1973, pp. 55, 60.

Chapter 6 Managing Goryōrin: Aspects of Practice

1 The following material on mensuration comes from Ōgane, 1973, pp. 150–3.

2 Ringyō hattatsushi, 1960, pp. 155–6.

3 This section is mainly summarized from Ringyō hattatsushi, 1960, pp. 471–5.

4 On the Tokugawa silvicultural heritage, see Totman, 1989, chapter 5.

5 According to Kuroda, 1966, p. 114, during 1926–37 the cost of hand-planting averaged 203 yen per hectare vs. 6 yen per hectare for natural reseeding.

6 To explain the terms "plant succession and continuous forest," they refer to the silvicultural truism that different plants flourish in different conditions and have differing lifespans. Especially where hillsides are steep, as in most of Japan's woodland, it is desirable after logging a site to have new growth appear quickly, its roots to hold the soil in place and its leaf cover to shelter the soil from hard rainfall. Such growth, mostly grasses and forbs, holds the ground surface intact for a few years while relatively fast-growing shrubs and deciduous trees establish themselves. These in turn provide the partial shade that slower growing conifers welcome. By skillful cutting and thinning, the forester helps these diverse forms of growth function optimally to achieve the longer-term production goal of a well-stocked stand of merchantable timber growing on stable terrain.

7 This material on Hokkaido planting comes from Ōgane, 1973, pp. 56, 60, 61, 62–3, 69–71.

8 Wakino, 1994, examines this topic as it relates to Kiso.

9 Ōgane, 1973, p. 61.

10 On Tokugawa-era logging technology see Totman, 1995. It treats briefly Kiso rafting and the Nishikori landing.

11 Ringyō hattatsushi, 1960, p. 469. By 1947, when its assets were turned over to the Forest Agency, the Household Ministry claimed to have 1,112 km of forest railroads, 6 km of tramways, and 979 km of vehicle roads. Hagino, 1996, p. 61.

12 This material on Kiso is assembled from Kuroda, 1966, p. 87; Wakino, 1994, pp. 2–3, and Nihon ringyō, 1997, pp. 417, 419, 421, 435, 441, 451.

13 On Otaru, see Hagino, 1990, pp. 151–2, and Hagino, 1993, pp. 118–22.

14 This annual rate implies roughly seventy to seventy-five days of logging activity. That figure, in turn, suggests how fully the Property Office relied on off-season work by local farmers.

15 Hōjō, 1983, pp. 476–7. Two years later Tokoro Mitsuo (1900-??), native of Matsumoto and enthusiastic student of matters forestal, published Tokoro,

1933, his lovely, masterful, fact-laden essay that memorialized Nishikori. That little essay, together with the monumental study of Edo-period silviculture, Tokugawa, 1941, were two of the works that first stirred my fascination with Tokugawa forest history.

16 This material on sawmills comes from Ringyō hattatsushi, 1960, pp. 478–9.

17 A generator for electric lighting would have permitted longer working days at the mill and may have given it a capacity that exceeded the Kiso Goryōrin's timber output.

18 Unless otherwise noted, this section is based on Ringyō hattatsushi, 1960, pp. 476–82.

19 Here, too, figures can be tricky. As noted in note 6 of Chapter 5, some records treat the final cut in a rotation block system as a selective cut. So a statistical decline in clearcutting can accompany a proliferation of plantation stands.

20 Hagino, 1993, p. 61, Table 7.

21 Kuroda, 1966, pp. 55, 88.

22 Hagino, 1996, pp. 46–7, discusses briefly 1930s developments in pulpwood production, in which chemists figured out how to make *akamatsu* usable for pulp.

23 See also Table 44 at Ringyō hattatsushi, 1960, p. 481. How this measure of increment (*seichōryō*) is established is unclear to me. On the ambiguities and complexities of such mensuration, see Daniel, 1979, pp. 304ff.

24 Hagino, 1990, esp. pp. 155–73, discusses timber exports.

25 In a technical sense timber from Karafuto was not an import because the area was part of the empire and its woodland was administered from Tokyo as *kokuyūrin*. In Korea and Taiwan, by contrast, local colonial administrations handled the woodland. Timber production on Karafuto soared during the twenties because an insect infestation was sweeping through the island's *ezomatsu* and *todomatsu* stands, and loggers were rushing to take out dead and dying stock while it was still merchantable. *Beizai*, or North American timber, mostly from the USA, was viewed as the great import menace. On that topic, see Hagino, 1990, pp. 192–258.

26 For an alternative view of Household self-assessment, see Bix, 2000, p. 552.

27 Kuroda, 1966, pp. 248–9, Table 2, provides the annual Goryōrin figures for income, expenses, balance, and land-sales income, 1889–1937. The dotted line in my Figure 6-4 combines income and land-sales figures, to bring Kuroda's numbers more into line with those of Hagino, 1996, who appears to have done so.

28 Goryōrin also had much better timber stands than *min'yūrin*, which served so many other purposes (in 1939, 407 *koku* of stumpage per *chōbu* compared to 206 for *min'yūrin*, as Table E-1 indicates). But during the war years private landholders harvested what they had much more aggressively (due to a combination of government pressure, local hardship, the profits to be made, and probably patriotism), in the process producing more lumber per hectare of woodland than either *kokuyūrin* or Goryōrin. Mitsui, 1986, p. 20, gives figures on saw-timber production for *min'yūrin* and combined *kokuyū/goryōrin*, 1930–60.

29 Borton, 1955, Appendix 4, p. 504.

Chapter 7 Goryōrin: An Assessment

1 The distinction I have in mind between agricultural and industrial societies is that of an economy sustained by current, domestic biomass energy output vs. an economy based on a global resource base and heavily dependent on fossil fuel energy. That theme is more fully developed in Totman, 2005.

2 This is not the fiscal picture painted by Bix, 2000, pp. 94–5, 144, 173, 552, who sees the imperial institution as a securely-wealthy colossus.

3 Goryōrin, like *kokuyūrin*, has been vigorously criticized by some scholars, particularly during the early post-1945 years, as part and parcel of an oppressive "absolutist" (*zettaishugi*) regime. Most of those critics, however, have not been scholars versed in matters silvicultural, and they have not been addressing ecological issues.

4 Government forest (*kokuyūrin* and Goryōrin combined) yielded on average about six million cubic meters of wood per year during the thirties, hit a wartime peak of 14.5 in 1943, fell below eight by 1950, and reached 14.6 in 1960. Mitsui, 1986, p. 20, Table I.

5 Albion, 1926, is the classic study of this topic.

6 James, 1981, p. 167, mentions a policy of forest enclosure during the nineteenth century that was designed to promote plantation forestry. However, the scale was minor: about 100,000 acres, compared to the millions of acres during the earlier enclosing for pasturage.

Part II

Chapter 8 The *Kan/Min* Division of Woodland in Early Meiji, 1871–76

1 By "Japan Proper" (or Naichi) I mean here the archipelago south of Hokkaido. This chapter does not treat that island, which at the time was a frontier territory, its woodland administered independently. Hokkaido did not contain Naichi's set of pre-industrial woodland arrangements, and its forest history developed quite differently from that to the south. It had more in common with the forest history of pre-1945 Karafuto, Taiwan, and Korea.

2 To cite a few of the key works in Japanese, a classic study of the tax reform is Fukushima, 1962, and a major collection of documents is Chiso kaisei, 1952–1957. Also valuable are Arimoto, 1968; Rin'ya seido, 1977; and the essay series Hōjō, 1988ff.

3 Noteworthy English-language sources on the 1870s land-tax reform include the following. The fullest general treatments are found in Beasley, 1972, pp. 390–400; Ike 1947, pp. 160–82; Nakamura, 1966, Appendix A, pp. 177–96; Niwa, 1966, pp. 447–71; Smith, 1955, pp. 382–99; and Yamamura, 1986, pp. 382–99. An older work that has valuable information is Rein 1889, pp. 9–12. Glimpses of land-tax reform as it applied in specific localities may be found in Chambliss, 1965, pp. 66–83; Kelly, 1985, pp. 163–4, 231–76; Smethurst, 1986, pp. 50–1, 136–8; and Wigen, 1995, pp. 184–8.

4 Different sources give different totals and breakdowns of acreage. The figures here come from Nakamura, 1966, Tables 1a, 1b, 1c, pp. 221–3. On p. 26 he calculates 4.47 million *chōbu* of arable in 1880, rising to 5.25 million by 1920. The

roughly two million *chōbu* of non-paddy-fields consisted primarily of dry fields for annual grain and vegetable production. However, about 10 percent (200,000 *chōbu*) were tea, mulberry, paper mulberry, lacquer, tallow, and fruit tree orchards, according to Rein, 1889, pp. 11–12.

5 These studies have focused on arable land in some cases because the authors were concerned with government tax income, for which arable was crucial and woodland nearly inconsequent. In other cases, the focus has been agriculture and its "modernization," for which issues relating to arable seemed central while the importance of non-arable was not apparent – perhaps because woodland products were so marginal to most of later-twentieth-century American agriculture.

6 In terms of biology, woodland (*sanrin gen'ya*, commonly shortened to *rin'ya*) ranged from pristine natural wilderness – which was marked by messiness, diversity of growth, and particularistic adaptation to soil, climate, and topography – all the way to grass monocultures fully subjected to human exploitation. Most woodland, however, contained some stumpage and a variegated understory of seedlings, brush, ferns and other forbs, viney growth, and grasses that ranged from towering bamboos to the most modest turf.

On the basis of its physical character, roughly half the woodland of mid-nineteenth-century Japan Proper qualified as *sanrin*, essentially meaning steep hillsides that were more-or-less fully stocked with trees. The other half, the *gen'ya*, included those high mountain surfaces, low hillsides, flood plains, and other more-or-less treeless areas dominated by brush, bracken, grasses, or lesser growth. The tidiness of the two labels notwithstanding, the biological distinction between them was in fact not at all clearcut, and vast amounts of poorly-forested land could plausibly be called one or the other, as a person chose. The persistence of these dyadic categories reflected the biology of woodland less than the uses people made of it – *sanrin* yielded timber; *gen'ya* yielded diverse other goods.

7 The contemporary boom in silk production, and hence in mulberry tree cultivation, added to tensions over woodland use in numerous areas, such as Yamanashi, as noted in Chapter 4.

8 The modern-day terms for "use rights" are *riyōken* and *yōekiken*. To my knowledge the best English-language study of *iriai* matters remains Wigen Lewis 1985.

9 "Multiple use" is a standard forester's term that highlights the diversity of woodland growth and its possible uses. "Shared use" is the term I use here because it emphasizes that more than one person or group has use rights on a given parcel of land.

10 The term "communal land" includes land controlled by villages as entities (*mura iriaichi*) and land controlled by associations of villagers (*kumiai iriaichi*). Rulers allotted a small portion of lord's forest to the court as imperial land (*goryōchi*) and to shrines and temples as fane land (*shajirin*), the latter process sometimes simply acknowledging a fane's existing claim to the land.

11 The *kan/min* distinction of the 1870s resembled – and may have been an early Meiji application of – the utterly ambiguous Anglophone "public/private" dyad. Meiji leaders seem initially to have equated it with the samurai/commoner (or

imperially titled/untitled?) dyad so that "public" became *kan'yū* or governmental, while "private," or *min'yū*, encompassed everything else. Since villages were administered by local commoners, not the vassals of shogun or daimyo, their property fell into the category of *min'yū*, along with that of all non-imperial fanes and non-titled householders.

12 Prior to the 1870s land-tax reform, some parcels of plantation forest and wooded areas used as swidden or for other more-or-less regular agricultural purposes were taxed. By one calculation, however, over 90 percent of the woodland in Japan Proper carried no regular tax at all. After the reform, arable land was taxed, on average, at the rate of 10.2 *yen* per *chōbu* whereas taxed woodland (i.e. *min'yūrin*) carried a rate of only 0.094 *yen* per *chōbu*. Figures from the table at Funakoshi, 1954, p. 5.

13 To say that Meiji leaders were ill-informed about woodland affairs is not to say that they were indifferent, as the massive documentation of exchanges between central and regional officialdom reveals. In 1879, moreover, as Tokyo was gearing up to take direct control of its sprawling woodland empire, "government leaders instructed the recently formed Forest Agency (Sanrinkyoku) to assemble documents that would aid in drafting legislation for a new forest system. Prefectural officials collected them, had copies transcribed, submitted the copies to Tokyo, and prepared reports detailing practices within their jurisdictions. As material flowed into the capital, agency scribes compiled it in a massive collection titled *Sanrin enkakushi* [Historical records of forestland]." Totman 1989, p. 195.

14 Since those who ousted the Tokugawa claimed to be restoring to the emperor the sacred authority he once supposedly enjoyed, they likely would have affirmed Emperor Shōwa's ancient claim to all the realm, as quoted in Chapter 1. But, in fact, any such claims had become romantic illusion centuries before 1870, and as a practical matter the only woodland the new Meiji leadership could plausibly claim as its own was the ohayashi of the fallen Tokugawa.

15 Hōjō, 1977, pp. 3–7, examines early use of the term *kanrin*. Today's term, *kokuyūrin*, was adopted in 1897, and it distinguished "national forest" from both *min'yūrin* and *goryōrin* (Imperial Household woodland). The latter had been formed from *kan'yūrin* during the preceding decade and was recombined with it in 1947.

16 Hōj 1988ff, part. 1, p. 103. Dajōkan *tasshi* No. 165 (1871.4.3). Note, however, that the notice was untraditional in two ways. It specified that the timber was important for building warships, and the trees that it named – *tsuki, kuri, kashi, matsu, buna*, and *shōji* (zelkova, chestnut, oak, pine, beech, and ?) – were not the ones (*sugi* and *hinoki* most centrally) found in Edo-period decrees of the same nature.

17 A Minbushō *tasshi* dated 1871.7. Translated from Hōjō, 1978, p. 160. Also given in Rin'ya seido, 1977, p. 6.

18 Hōjō, 1988 et seq., part 1, p. 104. Dajōkan *tasshi* No. 617 (1871.11.24).

19 Some officials saw the sale of *kanrin* parcels as a way to both use and assist disenfranchised samurai by returning them to productive work on the land, mainly in Tōhoku and Hokkaido. The effort, which continued into the 1880s,

did settle several thousand samurai households, but in terms of aiding the disenfranchised, it proved only a modest success. Many samurai simply sold the land they had received, pocketing the income; others tried to use it, failed to succeed, and sold out. Much more *kanrin* was sold directly to commoners, some of whom simply harvested the timber, others of whom tried to convert the land to arable or other uses, such as growing mulberry trees for silk production. On the samurai, see Harootunian, 1960, pp. 435–39.

20 Rin'ya seido, 1977, p. 22. Ōkurashō *tasshi* No. 76 (1872.6.15), fourth clause.

21 The following summary reflects materials in several sources, including Ringyō hattatsushi, 1960, pp. 192–5; Hōjō, 1988ff, part 4, pp. 128–52, and Arimoto, 1968, pp. 313–15. On Edo-period *warichi*, see Brown, 1997, pp. 421–44.

22 The government's concurrent attempts to consolidate villages complicated its task of securing collaboration by local officials and villagers. However, that consolidation encountered so much opposition that it was nearly abandoned by Meiji leaders in February 1875, and it made little progress before the late 1880s (there were roughly 80,000 local administrative units in 1870, 70,000 in 1887, but fewer than 16,000 by 1889), so we disregard it here. For an introduction in English to the topic of village consolidation, see Steiner, 1968, pp. 221–36. Also see Umegaki, 1988, pp. 130–48, and Steele, 1989, pp. 729–48.

23 Rin'ya seido, 1977, pp. 12–15. Ōkurashō *tasshi* No.126 (1872.9.4), an addendum to Ōkurashō *tsutatsu* No.25 (1872.2.24).

24 Reflecting the Finance Ministry's perception of woodland as a fiscal resource, and in preparation for the anticipated property-tax inflow, in 1872.10 the Ministry transferred forest matters from its agriculture section to its taxation department, which was a larger and more influential office. Nishio, 1988, p. 30.

25 Hōjō, 1988 et seq., part 5, pp. 335–66, treats prefectural responses to Tokyo's *kōyūchi* guidelines.

26 Ringyō hattatsushi 1960, p. 143. This ruling on *muramochi rin'ya* also complicated later efforts at village consolidation because it gave many hamlets legal title to their woodland. Later, when the government tried to consolidate these hamlets (formerly called *mura* but relabeled *buraku*) into larger towns and villages, their residents were for decades able to retain hamlet control of their woodland, which undercut government efforts to use that woodland to pay for local schools, clinics, and other municipal facilities. A recent look at the role of woodland in village consolidation is Hōjō, 1988 et seq., Parts 7–8, 10–12.

27 Hōjō, 1988 et seq., part 6, pp. 55–6; Rin'ya seido, 1977, pp. 29–30; Arimoto, 1968, pp. 315–16. Dajōkan *fukoku* No.114 (March 25, 1873). In terms of acreage the three main land categories were "government," "communal," and "private" land (*kan'yūchi, kōyūchi*, and *shiyūchi*). The other five were Imperial Household land, shrine land, central and regional government office-use land, and tax-exempt municipal land. Tax-exempt municipal land included roads, dikes, public signpost sites, execution grounds, cemeteries, and village fane lands.

28 For a brief survey of the *kōyūchi* issue during 1872–75, see Ringyō hattatsushi, 1960, pp. 137–56. Also Kasai, 1984, pp. 1–12. Kasai emphasizes the importance of prefectural correspondence in acquainting Tokyo officials with *kōyūchi* and related problems and thus shaping forest policy.

29 Rin'ya seido 1977, pp. 30–4, reproduces Dajōkan *fukoku* No. 272 and the detailed Ōkurashō *tasshi* for its implementation.

30 Because reformers had no adequate basis for establishing the market value of land, they fell back on a modified version of the Tokugawa method for setting tax levels. As explained in the instructions of July 28, after identifying the ownership, acreage, and soil quality of specific parcels of land, they were to obtain from the owner estimates of his land's normal productive capacity and his main costs of production. Then, using a predetermined formula, assessors could establish a hypothetical cash value for the owner's net annual farm income. Finally, treating that number as equivalent to the interest payable on a mortgage, they used local mortgage interest rates to calculate the full value of that fictive debt. The resulting figure became the assessed value of those parcels. For further detail, see Nakamura, 1966, Appendix A, and Yamamura, 1986.

31 Nishio, 1988, p. 36.

32 It perhaps should be noted that resistance to the civil-war victors had been strongest in the northeast, while the southwest was whence those victors came.

33 I write "more completely fungible" because in one of the innumerable compromises made along the way, for several years villagers were allowed, under specified circumstances, to pay their tax in kind. See Mayet 1893, p. 228.

34 Hōjō, 1988 et seq., part 2, p. 206. Dajōkan *fukoku* No. 257.

35 Hōjō, 1988 et seq., part 2, pp. 206–209. Ōkurashō *tasshi* No.134 .

36 According to the form that accompanied Notice No. 134, officials were to arrange their information on government woodland in terms of two basic categories: parcels to be retained and those to be sold. Within each category, three types of property were to be listed: parcels identified simply as *kanrin*; parcels on which *iriai*-type use rights were established; and parcels of *kōyūchi*. For each simple *kanrin* site in both the "keep" and "sell" categories, prefectural assessors were to identify the location; specify its boundaries in terms of abutting hills, rivers, arable, or waste; indicate its topography and acreage; and report any records of its past use in furnishing timber for castles, mansions, or other construction, or its status as "forest reserve" (*tomeyama*), or any other evidence of its silvicultural distinction. In the second sub-category, that of *iriai*-burdened *kanrin*, officials should record those sites on which people had removed undergrowth for specified fees, with the relevant village use-arrangements indicated. *Kōyūchi* was similarly to be reported, with information on such fees as users paid, including those areas on which taxes were and were not levied.

37 One suspects, too, that the form accompanying Notice No. 134 may have been designed to identify those localities with the most valuable woodland and also those prefectures in which officials seemed most committed to securing government control of good woodland or most ready to accommodate local interests – the better, perhaps, to assure consistent pursuit of Tokyo's priorities.

38 Iwata, 1964, p. 161. Nishio, 1988, pp. 30, 34, stresses the influence of Inoue, who served as "vice-chief" of the Finance Ministry (Ōkurashō *taiho*) from July 1871 to May 1873.

39 Arimoto, 1968, pp. 301–304.

40 Ōkubo returned on May 20, Kido on July 23, and Iwakura on September 13. Because of the raging disagreements over Korea policy, Ōkubo avoided overt participation in affairs until after Iwakura's return. But one suspects that he made his views known where doing so seemed useful, and on forestry matters he emerged as a strong proponent of government woodland management.

41 Nihon shiseki kyōkai, 1968, Vol. VI, pp. 363–6. Sanjō's title was Dajō daijin. The supplemental essays, if ever written, are not included. Being Ōkubo's fullest expression of his views on matters silvicultural, this memorandum is widely cited, as in Nishio, 1988, pp. 37–8; Ringyō hattatsushi, 1960, pp. 272–8; Nagatsuma, 1987, pp. 1–2; Narita, 1996, p. 53.

42 Most notably, perhaps, in August 1875 this interest in silviculture was strengthened when Matsuno Hazama – who had been studying German forestry in Berlin and had talked to Kido and Ōkubo about the topic when they visited the city with the Iwakura Mission – returned to Tokyo and at Ōkubo's behest took a post in the new Forestry Department. There he played a key role in developing woodland-surveying procedures and advising on the criteria for *kan/min* division, with the explicit goal of bringing as much good timberland as possible under government control. Nishio, 1988, pp. 42–4. Matsuno's given name can also be read as Kan.

43 Woodland sales were not totally abandoned. Indeed, renewed efforts were made by officials to sell land to ex-samurai as a way to reduce their anger at the regime. But while the topic is fascinating, the effort was of sufficiently modest consequence for woodland that we omit it here. See footnote No.19 above.

44 Rin'ya seido, 1977, pp. 83–7, unnumbered Naimushō *ketsugi*. Ringyō hattatsushi, 1960, pp. 57–60.

45 Rin'ya seido, 1977, p. 57. This notice, No. 1874–9, March 7, 1874, came out over the seal of Kido Kōin. At the time, he was serving in Ōkubo's place while the latter dealt with the Saga Rebellion.

46 Niwa, 1988, pp. 56–64. Where *yakibata* was taxed, Niwa reports, its use tended to become uneconomic and villagers mostly ceased tilling it, causing a sustained drop in the extent of swidden culture.

47 Rin'ya seido, 1977, pp. 20–30, treats these matters. Also Hōjō, 1988 et seq., part 6, pp. 60–71, looks at prefectural queries of this period.

48 Ringyō hattatsushi, 1960, pp. 150–1, Dajōkan *fukoku* No.120, November 7, 1874; Rin'ya seido, 1977, pp. 63–5.

49 Kasai, 1987, pp. 2–4, looks at prefecture-Tokyo exchanges on issues in *kan/min* division, 1874–76.

50 Nishio, 1988, pp. 36–7.

51 Ringyō hattatsushi, 1960, p. 44, Naimushō *tasshi* No.83. Ōkubo issued this Notice in his role as Interior Minister.

52 Rin'ya seido, 1977, p. 71, *tasshi* No.3, June 22, 1875.

53 Hōjō, 1977, pp. 48–52, 61–84.

54 Kasai, 1987, pp. 3–4, reviews this topic.

55 Rin'ya seido, 1977, pp. 82–3, unnumbered *giteisho* of January 29, 1876. Also Hōjō, 1977, pp. 39–41.

56 Tateishi, 1976, pp. 156–8.

57 Shōnai is treated in Tateishi, 1976, pp. 159–78; Tateishi, 1980, pp. 115–35, and

Tateishi, 1983, pp. 113–38. In remarkable diagrams on pp. 116 and 123, Tateishi, 1983 reveals the intricacy of dunes property and foliage boundaries in Shōnai.

58 Quoted in Watanabe, 1988, p. 28.

59 Watanabe, 1988, pp. 23–31.

60 Narita, 1996, pp. 47–63, examines this Watarai matter.

61 Tokyo's seizure of fane (temple and shrine) lands gave the government another 114,413 *chōbu* of *kanrin*, leaving the fanes with a registered 25,608 *chōbu* of land within their compound walls. (Ringyō hattatsushi 1960, p. 237.) Fane woodland constituted a sore point all by itself, and as the documents clearly reveal, it added one more complication to woodland reform, but for the sake of brevity, I have disregarded it in this chapter.

62 Narita, 1996, p. 56.

63 Precisely because the "Kiso Incident" became such a high-stakes contestation, recollections and interpretations are harshly at odds, and this summary sidesteps much of the conflict, attempting to provide a concise narrative based on what appear to be judicious scholarly reconstructions of the affair.

64 The other two *birin* were domanial forests found in today's Akita and Aomori prefectures. Both are too cold for *hinoki*, and *sugi* flourish only on lower elevations of Akita.

65 Machida, 1978, p. 176.

66 Machida, 1978, pp. 177–9.

67 Ōshima, 1981, pp. 220–1.

68 Machida, 1978, pp. 182–4; Ōshima, 1981, pp. 218–19.

69 Hōjō, 1977, pp. 39–41.

70 Machida, 1978, pp. 184–6.

71 Machida, 1978, pp. 186–8.

72 Machida, 1978, pp. 187–91; Ōshima, 1980a, pp. 156–9; Ōshima, 1982, pp. 152–6.

73 Machida, 1978, pp. 191–3. Ōshima, 1982, p. 157.

74 Besides the 220,000 *chōbu* of *akeyama*, the Kiso area included an estimated 160,000 *chōbu* of *ohayashi* that did not carry *iriai* rights, plus perhaps 8,000 *chōbu* of arable.

75 Machida, 1978, pp. 193–9. Ōshima, 1980a, pp. 164–74, and Ōshima, 1982, pp. 157–63, examine the survey-related correspondence of 1881–83.

76 On the Kiso Incident more broadly, a number of fine works exist, other than the journal articles cited here. They include Aoki, 1968; Hōjō, 1983; Machida, 1982. On later developments, and a guide to source materials, see Shimizu 1986, pp. 26–37. The Incident also is treated, somewhat loosely, in Shimazaki Tōson's sprawling novel *Yoakemae* (translated into English by William E. Naff as *Before the Dawn*, University of Hawai'i Press, 1987). See especially pp. 564–8, 575–80, 688–90. Tōson's treatment is examined in Hōjō 1998 and earlier works cited therein.

77 Ōshima, 1980a, pp. 159–60. Ōshima, 1980b, pp. 177–206, studies the logging activity of Saruzawa village in Kiso during 1880–83 and the boundary dispute associated with that work.

78 Ringyō hattatsushi, 1960, pp. 61, 98, 245. Figures differ from source to source,

the diversity reflecting both the complexity of record keeping and the uncertain quality of the figures.

Appendices

Appendix B

1 A concise discussion of this topic is in Titus, 1974, pp. 66–7, which also adumbrates the Household's accounting system.

2 Figures from Titus, 1974, p. 68. Note, however, the much higher Goryōrin average annual balance figure of 23,948,000 yen for 1939–43 in Table E-13. (It is calculated from Hagino, 1996, p. 60, Table 6, which gives the figure for 1940 as 21,840,000 yen, and which seems to include any land sales made that year). Titus's 9,000,000 yen estimate is a reasonable projection from the 1930s figures found in Kuroda, 1966, at p. 249, Table 2, but it doesn't account for inflation and accelerated logging of the late thirties.

3 Kuroda, 1966, pp. 15, 30, 71.

4 Bureau of Imperial Forest figures from Hagino, 1996, p. 66, Table 9; Household totals from Nezu, 1961, p. 59. Nezu notes that with inflation the total was revalued to some 4.45 billion yen in March 1946, and he cites Tanaka Sōgorō's figure of 6.0 billion from *Tennō no kenkyū*, p. 299. In his later, shorter work, Nezu, 1974, p. 266, gives wildly higher figures but offers no supporting documentation for them. Kuroda, 1966, p. 253, Table 6–2, gives figures that total 3,176,787 yen. Most notably, he omits the "Fine Arts" figure.

5 Lest one attach too much credence to these figures, however, it is worth quoting Titus, 1974, pp. 68, footnote 23: "The biographers of Ishiwatari Sōtarō, the last Imperial Household minister, claim that there had never been a complete estimate of imperial house properties until the Occupation Authorities demanded such a valuation after the war. The value finally arrived at by Ishiwatari was 1.6 billion yen." Nezu, 1974, pp. 265–6, cites Ishiwatari's figures.

Appendix C

1 Itō, 1906, p. 180. On the provenance of Itō's *Commentaries*, see Siemes, 1968, p. 16.

2 Itō, 1906, pp. 155–6. See Titus, 1974, pp. 43–4, for a concise discussion of this issue.

3 Itō, 1906, p. 155. Also quoted in Borton, 1955, Appendix 4, p. 506.

4 Itō, 1906, p. 139. The following comment is from p. 140. Another translation is in Siemes, 1968, p. 231.

5 The two quotations come from Itō, 1906, pp. 111–12, 114.

6 Itō, 1906, pp. 54, 55.

Bibliographical Essay

My search of bibliographical sources revealed no English-language works on Japan's Imperial Forest, so this essay addresses only works in Japanese. Even there the corpus is modest, and key works are dated and difficult to find. Nevertheless, treating Goryōrin independently does have advantages. It permits one to examine several issues in Japan's forest history on a more modest – and hence more manageable – scale than would be required for a wide-bodied study that encompassed Goryōrin, *kokuyūrin*, and the various forms of *min'yūrin*.

Moreover, the dissolution of Imperial Forest in 1947 enables one to sidestep the history of Japan's forests during the half century to 2000, a topic that is exceptionally fascinating and fully deserving of attention but that introduces major new issues and developments, as hinted at in Chapter 7. We should, however, note the recent appearance of a valuable collection of essays on aspects of recent Japanese forestry and the lumber industry: Yoshiya Iwai, *Forestry and the Forest Industry in Japan*.

Several books and articles on the history of Imperial Forest deserve specific mention, both for their own merit and to exemplify the range of pertinent work on this topic.

1. GENERAL WORKS ON PREWAR FOREST HISTORY

Works of this sort usually allot some space to Goryōrin. The great progenitor of the genre, *Meiji ringyō shiyō* (2 vols.) and its addendum *Meiji ringyō shiyō kōshū*, by Matsunami Hidemi (see the bibliography for full citations), was the work of a former head of the government's Forest Agency, and he devoted most of his attention to *kokuyūrin*, giving only modest attention to Goryōrin and *min'yūrin*. The next major official treatment of the pre-1940s history, *Nihon ringyō hattatsushi (jō)*, compiled by Ringyō hattatsushi chōsakai, gave Goryōrin and its vicissitudes substantially fuller attention, and it proved especially helpful in this project. Its successor volume, on the other hand, *Nihon ringyō hattatsushi*, compiled by Dai Nihon sanrinkai, which follows the forest story from the 1920s through World War II, contains very little on Goryōrin. The much shorter survey by Tsutsui Michio, *Nihon rinseishi kenkyū josetsu*, offers concise summaries of main aspects of Goryōrin developments. And among the several major studies by Hagino Toshio, his *Nihon gendai rinsei no sengō katei*, while centrally a study of post-1945 developments, provides valuable data on trends in Goryōrin from the 1930s until its dissolution.

Beyond these general studies, there is a modest literature that treats Goryōrin exclusively. The topic also shows up as seems useful in works on regional forestry, or in works focused on particular problems, such as *iriaichi*. A few studies appeared during Goryōrin's heyday; most have appeared since then.

2. PREWAR STUDIES OF GORYŌRIN

The single most valuable prewar study, to which I have not had access, is the data-packed volume, *Teishitsu rin'yakyoku gojūnenshi*, compiled by Wada Kunijirō and published by the Bureau of Imperial Forests for its fiftieth anniversary in 1939. Fortunately, major scholarly studies have relied heavily on that work for basic statistics down to 1938, and I have used them here. Wada and the Bureau also produced the slightly earlier *Meiji Taishō goryō jigyōshi*. It is a richly technical study that covers some of the same material as the *Gojūnenshi*. Also of prewar vintage is the monthly journal *Goryōrin*, which appeared during the thirties. As its name suggests, it contains numerous technical studies, research reports, and interpretive articles dealing with the operations, situation, and history of Imperial Forest.

3. POSTWAR STUDIES OF GORYŌRIN

Since the Pacific War a few book-length scholarly works on Goryōrin have appeared. Two of the most solid focus on the celebrated disputes involving Imperial Forest in Kiso, whose background received fictional treatment in Shimazaki Tōson's sprawling novel *Yoakemae* (translated into English by William E. Naff as *Before the Dawn*). Aoki Keiichirō's *(Shiryō) Kiso goryōrin jiken kōshōroku* is a detailed, chronologically organized narrative of documents and developments for the years 1898–1905. More recently, the same "Kiso Incident" has been examined more analytically by Machida Masami in his *Kiso goryōrin jiken*. Machida's endnotes and postscript provide guidance to sources and earlier studies, including his own.

Works on regional forestry have dealt with Goryōrin where it was an important part of the whole, most notably in Hokkaido and Kiso. *Hokkaidō ringyō gijutsu hattatsu shiron*, a study of Hokkaido forestry edited by Ōgane Eiji et al., contains three short, solid essays on Goryōrin operations, primarily as seen in records of the Sanru operations area. Its footnotes guide one to source materials, including articles in *Goryōrin*. On Kiso see, in addition to the Aoki and Machida volumes cited above, Hōjō Hiroshi's *Meiji kokka no rin'ya shoyū to sonraku kōzō: Nagano-ken Kiso kokuyūrin no sonzai keitai*. Also, in an article published in 1986, "Kiso goryōrin jiken no ichi kōsatsu," Shimizu Yūzō examines later lawsuits in the case and identifies salient sources in his footnotes.

A number of scholarly books and articles have touched on Goryōrin as it related to some broader topic or have dealt with select topics in Goryōrin history. One notable example is Kuroda Hisata, *Tennōke no zaisan*, a study of the imperial fisc that devotes substantial space to Goryōrin. A recent article that I have not seen is

Shima Yoshitaka, "Meiji 23 nen no seden goryō chokujō ni tsuite," which examines circumstances surrounding the original decree establishing hereditary Imperial Forest in 1890. An article that came to my attention after this study went to press is Ōsawa Satoru, "Goryōchi keisei katei no ichi danshō," which analyzes an 1887 survey of woodland in Gunma Prefecture, as an example of the procedures employed in selecting woodland parcels for inclusion in Goryōrin.

The broader question of *iriai* and Goryōrin-village relations is treated by a number of authors. Ōshima Mario's article, "Meiji-ki no sanson ni okeru rin'ya riyō to kyōdōtai kihan," examines the *iriai* situation on Goryōrin in Yamanashi Prefecture near the north slope of Mt. Fuji. Two other examples are works by the prolific Hōjō Hiroshi, the earlier one his introduction to issues in Meiji forest history, *Meiji ki ni okeru sanrin mondai no kenkyū*, which provides a concise treatment of the Yamanashi situation, together with a few pertinent documents. More recent is his detailed *Rin'ya iriai no shiteki kenkyū (jō)*, also focusing on Yamanashi, which examines the issue of *iriai* in Goryōrin as well as *kokuyūrin*. In conjunction with his study of Yamanashi, Hōjō also supervised, under the auspices of the Rin'ya seido kenkyūkai, publication of *Kindai rin'ya seido shiryōshū*, a fine set of documents on Meiji-era forestry, mainly relating to Yamanashi.

Finally, one must note three reference works. Two older but useful works in English are *Important Trees of Japan* and *Glossary of Japanese Forestry Terms*, both issued by the Natural Resources Section of the Supreme Commander for the Allied Powers. A recent and exceedingly useful volume is *(Sōgō nenpyō) Nihon no mori to ki to hito no rekishi*, a chronology of forestry compiled and published by Nihon ringyō chōsakai. It contains side-by-side chronologies of general history, government forest policy, forest operations, and forest industry from the Old Stone Age to the end of 1989. Not only is it full of interesting detail, but it also enables one to resolve inconsistencies of dating that appear in the scholarly corpus.

Bibliography of Works Cited

This Bibliography is organized in these sections:

A. ON GORYŌRIN
   1. Japanese-language sources
   2. English-language sources on Japan
   3. English-language sources on other topics

B. ON *KAN/MIN* DIVISION
   1. English-language sources
   2. Japanese-language sources

Each entry is preceded by the identifier used in the end notes – such as "Aoki, 1968" – and the works are alphabetized accordingly. Authorial names are then given as they appear in the works: surnames generally appear last in English-language works; but first in Japanese-language works.

ABBREVIATIONS:

CUP = Columbia University Press	UCP = University of California Press
HUP = Harvard University Press	UHP = Univesity of Hawai'i Press
PUP = Princeton University Press	UTP = University of Tokyo Press

RK = *Ringyō keizai*
TRK KK = *Tokugawa rinseishi kenkyūjo kenkyū kiyō*

A. ON GORYŌRIN

1. Japanese-language sources

Dai Nihon, 1983. Dai Nihon sanrinkai, comp., *Nihon ringyō hattatsushi* [A history of the development of forest industry in Japan] (Tokyo: Dai Nihon sanrinkai, 1983).

Goryōrin [Imperial Forest] (Tokyo: Rin'yakai, monthly, esp. 1930s).

Hagino, 1990. Hagino Toshio, *Nihon kindai rinsei no hattatsu katei: sono jisshōteki kenkyū* [The process of development of Japan's modern forest system: a study with substantiating evidence] (Tokyo: Nihon ringyō chōsakai, 1990).

Hagino, 1993. ______, *Nihon gendai rinsei no gekidō katei: kyōkō, jūgonen sensōki no jisshō* [The process of upheaval in Japan's present-day forest system: evidence from the panic and the fifteen-year war period] (Tokyo: Nihon ringyō chōsakai, 1993).

Hagino, 1996. ______, *Nihon gendai rinsei no sengō katei* [The postwar course of Japan's present-day forest system] (Tokyo: Nihon ringyō chōsakai, 1996).

Hōjō, 1965. Hōjō Hiroshi, *Meiji-ki ni okeru sanrin mondai no kenkyū* [A study of Meiji-era forest problems] (Tokyo: Keiō shoten, 1965).

Hōjō, 1968. ______, "Meiji chūki ni okeru rin'ya iriai no sonzai keitai (ge)" [The character of field and forest commonland in middle Meiji (part 2)] *TRK KK* 42 (1968): 35–100.

Kuroda, 1966. Kuroda Hisata, *Tennōke no zaisan* [The property of the imperial household] (Tokyo: San'ichi shobō, 1966).

Matsunami, 1919. Matsunami Hidemi, *Meiji ringyō shiyō* (2 vols.) [The essentials of Meiji forest history] (Tokyo: Hara shobō, 1990; orig. publ. 1919).

Matsunami, 1924. ______, *Meiji ringyō shiyō kōshū* [An addendum to 'The essentials of Meiji forest history'] (Tokyo: Hara shobō, 1990; orig. publ. 1924).

Mitsui, 1986. Mitsui Shōji, "Senjiki ni okeru sozai seisan seisaku no tenkai to shinrin kumiai/shinrin shoyūsha no taiō," [The development of wartime timber-production policy and the opposition of woodland guilds and woodland owners] *RK* 457 (39–11) (Nov. 1986): 15–32.

Mitsui, 1994. ______, "Kindaiteki rin'ya shoyū seisaku no tenkai hen'yō katei" [Evolution and change in modern woodland ownership] *RK* 549 (47–7) (July 1994): 7–19.

Nezu, 1961. Nezu Masashi, *Shōchō teikoku no tanjō* [Birth of the symbolic monarchy] (Vol. II of *Tennō Shōwaki*) [A record of the Shōwa period] (Tokyo: Shiseidō, 1961).

Nezu, 1974. ______, *Tennō to Shōwashi* [The emperor and Shōwa history] (Tokyo: San'ichi shobō, 1974).

Nihon ringyō, 1997. Nihon ringyō chōsakai (comp.) (with planning and editorial supervision by Kokudo ryokka suishin kikō), *(Sōgō nenpyō) Nihon no mori to ki to hito no rekishi* [The History of Japan's forests, trees, and people: a chronology] (Tokyo: Nihon ringyō chōsakai, 1997).

Ōgane, 1973. Ōgane Eiji et al., eds., *Hokkaidō ringyō gijutsu hattatsu shiron* [Essays on the development of forestry technique in Hokkaido] (Sapporo: Hokkaidō daigaku tōshō kankōkai, 1973).

Ōsawa Satoru, "Goryōchi keisei katei no ichi danshō: Ōki Makibi no Gunma-ken junkai to fukumeisho" [A document on the forming of Imperial Lands: Ōki Makibi's survey and report on Gunma Prefecture], *Kaetsu daigaku kenkyū ronshū*, 48 (Oct. 2005): 21–42.

Ōshima, 1976. Ōshima Mario, "Meiji-ki no sanson ni okeru rin'ya riyō to kyōdōtai kihan" [The uses of woodland and the norms of the community in Meiji-era mountain villages] *TRK KK* 50 (1976): 120–53.

Ringyō hattatsushi, 1958. Ringyō hattatsushi chōsakai, comp., *Nihon ringyō hattatsushi (jō)* [A history of the development of the forest industry in Japan (vol. 1)] (Tokyo: Rin'yachō, 1958).

Rin'ya seido, 1977. Rin'ya seido kenkyūkai (Hōjō Hiroshi), comp., *Kindai rin'ya seido shiryōshū* [Documents on the modern forest system] (Tokyo: Ochanomizu shobō, 1977).

Shima, 1993. Shima Yoshitaka, "Meiji 23 nen no seden goryō chokujō ni tsuite" [Regarding the 1890 decree on hereditary imperial forest] *Waseda jinbun shizen kagaku kenkyū* 44 (1993): 1–39.

Teishitsu rin'yakyoku, 1939. Teishitsu rin'yakyoku (Wada Kunijirō), comp., *Teishitsu rin'yakyoku gojūnenshi* [Fifty years of the Bureau of Imperial Forests] (Tokyo: Teishitsu rin'yakyoku, 1939).

Tokoro, 1933. Tokoro Mitsuo, "Nishikori tsunaba ni tsuite: kuchie kaisetsu" [The log boom at Nishikori: an introductory explanation] *Shakai keizai shigaku* 2–12 (1933): 101–118.

Tokugawa, 1941. Tokugawa Muneyoshi, *Edo jidai ni okeru zōrin gijutsu no shiteki kenkyū* [An historical study of afforestation techniques in the Edo period] (Tokyo: Nishigahara kankōkai, 1941).

Tokyo kyōiku, 1955. Tokyo kyōiku daigaku nōgakubu ringakuryō, comp., *Jitsuyō ringyō keisan hikkei* [Manual of applied forest mensuration] (Tokyo: Shōkōdō, 1955).

Tsutsui, 1978. Tsutsui Michio, *Nihon rinseishi kenkyū josetsu* [An introduction to the study of forest administration in Japan] (Tokyo: Tokyo daigaku shuppankai, 1978).

Tsutsui, 1987. ______, *Nihon rinsei no keifu* [A genealogy of Japan's forest system] (Tokyo: Chikyūsha, 1987).

Wada, 1935. Wada Kunijirō, *Meiji Taishō goryō jigyōshi* [A history of imperial forest operations in the Meiji-Taishō eras] (Tokyo: Teishitsu rin'yakyoku, 1935).

Wakino, 1994. Wakino Hiroshi, "Kiso shinrin tetsudō michibiki to Kiso-shiki batsuboku unzaihō" [Introduction of the Kiso forest railway and the methods of "Kiso style" felling and hauling] *RK* 547 (47–5) (May 1994): 1–10.

2. English-language sources on Japan

Akita, 1967. George Akita, *Foundations of Constitutional Government in Modern Japan 1868–1900* (Cambridge, MA: HUP, 1967).

Banno, 1992. Junji Banno (J. A. A. Stockwin, tr.), *The Establishment of the Japanese Constitutional System* (London: Routledge, 1992).

Beasley, 1963. William G. Beasley, *The Modern History of Japan* (London: Frederick A. Praeger, 1963).

Beckmann, 1957. George M. Beckmann, *The Making of the Meiji Constitution* (Lawrence: University of Kansas Press, 1957).

Bergamini, 1971. David Bergamini, *Japan's Imperial Conspiracy* (NY: William Morrow & Co., 1971).

Bix, 2000. Herbert P. Bix, *Hirohito and the Making of Modern Japan* (NY: Harper Collins, 2000).

Borton, 1955. Hugh Borton, *Japan's Modern Century* (NY: The Ronald Press, 1955).

Bowen, 1980. Roger Bowen, *Rebellion and Democracy in Meiji Japan: A Study of Commoners in the Popular Rights Movement* (Berkeley: UCP, 1980).

Butler, 2002. Lee Butler, *Emperor and Aristocracy in Japan, 1467–1680* (Cambridge, MA: Harvard University Asia Center, 2002).

Center, 1972. Center for East Asian Cultural Studies, comp., *Meiji Japan Through Contemporary Sources, Volume Three 1869–1894* (Tokyo: The Center for East Asian Cultural Studies, 1972).

Ericson, 1996. Steven J. Ericson, *The Sound of the Whistle, Railroads and the State in Meiji Japan* (Cambridge, MA: HUP, 1996).

Fujitani, 1998. Takashi Fujitani, *Splendid Monarchy, Power and Pageantry in Modern Japan* (Berkeley: UCP, 1998).

Hall, 1966. John Whitney Hall, *Government and Local Power in Japan, 500–1700* (Princeton: PUP, 1966).

Itō, 1906. Itō Hirobumi, *Commentaries on the Constitution of the Empire of Japan*, 2d ed. (Tokyo: Chūō daigaku, 1906).

Iwai, 2002. Yoshiya Iwai, *Forestry and the Forest Industry in Japan* (Vancouver: UBC Press, 2002).

Jansen, 1986. Marius B. Jansen and Gilbert Rozman, eds., *Japan in Transition from Tokugawa to Meiji* (Princeton: PUP, 1986).

Kido, 1983. Kido Takayoshi (Sidney Devere Brown and Akiko Hirota, tr), *The Diary of Kido Takayoshi* (3 vols.) (Tokyo: UTP, 1983).

Large, 1992. Stephen S. Large, *Emperor Hirohito and Shōwa Japan, A Political Biography* (London: Routledge, 1992).

Lebra, 1973. Joyce C. Lebra, *Ōkuma Shigenobu, Statesman of Meiji Japan* (Canberra: Australian National University Press, 1973).

Nakamura, 1966. James I. Nakamura, *Agricultural Production and the Economic Development of Japan 1873–1922* (Princeton: PUP, 1966).

Natural Resources, 1949. Natural Resources Section, *Important Trees of Japan* (Report No. 119) (Tokyo: Supreme Commander for the Allied Powers, 1949).

Natural Resources, 1950. _______, *Glossary of Japanese Forestry Terms* (Report No. 131) (Tokyo: Supreme Commander for the Allied Powers, 1950).

Seagrave, 1999. Sterling and Peggy Seagrave, *The Yamato Dynasty* (NY: Broadway Books, 1999).

Shimazaki, 1987. Shimazaki Tōson (William E. Naff, tr.), *Before the Dawn* (Honolulu: UHP, 1987).

Siemes, 1968. Johannes Siemes, *Herman Roesler and the Making of the Meiji State* (Tokyo: Sophia University, 1968).

Smethurst, 1986. Richard J. Smethurst, *Agricultural Development and Tenancy Disputes in Japan, 1870–1940* (Princeton: PUP, 1986).

Titus, 1974. David Anson Titus, *Palace and Politics in Prewar Japan* (NY: CUP, 1974).

Totman, 1989. Conrad Totman, *The Green Archipelago, Forestry in Pre-industrial Japan* (Athens: Ohio University Press, 1998, orig. publ. Berkeley: UCP, 1989).

Totman, 1995. _______, *The Lumber Industry in Early Modern Japan* (Honolulu: UHP, 1995).

Totman, 2004. _______, *Pre-industrial Korea and Japan in Environmental Perspective* (Leiden: Brill, 2004).

Totman, 2005. _______, *A History of Japan* (Oxford: Blackwell Publishers, 2005).

Tsunoda, 1958. Ryusaku Tsunoda et al., comps., *Sources of the Japanese Tradition* (NY: CUP, 1958).

Webb, 1968. Herschel Webb, *The Japanese Imperial Institution in the Tokugawa Period* (NY: CUP, 1968).

Wetzler, 1998. Peter Wetzler, *Hirohito and War, Imperial Tradition and Military Decision Making in Prewar Japan* (Honolulu: UHP, 1998).

Wigen, 1995. Kären Wigen, *The Making of a Japanese Periphery, 1750–1920* (Berkeley: UCP, 1995).

Wigen, 2000. ______, "Teaching About Home: Geography at Work in the Prewar Nagano Classroom," *Journal of Asian Studies* 59–3 (Aug. 2000): 550–574.

3. English-language sources on other topics

Albion, 1926. R. G. Albion, *Forests and Sea Power: The Timber Problems of the Royal Navy, 1652–1862* (Cambridge, MA: HUP, 1926).

Anderson, 1967. Mark Zoudin Anderson, *History of Scottish Forestry* (2 vols.) (London: Nelson, 1967).

Bambach, 1980. Richard K. Bambach et al., "Before Pangaea: The Geographies of the Paleozoic World," in Brian J. Skinner, ed., *Earth's History, Structure and Materials* (Los Altos, CA: William Kaufmann, Inc., 1980), pp. 86–98.

Daniel, 1979. Theodore W. Daniel, John A. Helms, and Frederick S. Baker, *Principles of Silviculture*, 2d ed. (NY: McGraw Hill, 1979).

Dietz, 1970. Robert S Dietz. and John C. Holden, "The Breakup of Pangaea," in *Scientific American* 223–4 (1970): 30–41.

Havins, 1976. Peter J. Neville Havins, *Forests of England* (London: Robert Hale & Co., 1976).

Heske, 1938. Franz Heske, *German Forestry* (New Haven: Yale University Press, 1938), p. 52.

James, 1981. N. D. G. James, *A History of English Forestry* (Oxford: Basil Blackwell, 1981).

Linnard, 1982. William Linnard, *Welsh Woods and Forests: History and Utilization* (Cardiff: National Museum of Wales, 1982).

McCracken, 1971. Eileen McCracken, *The Irish Woods Since Tudor Times: Distribution and Exploitation* (Newton Abbot: David & Charles, 1971).

McEvedy, 1978. Colin McEvedy and Richard Jones, *Atlas of World Population History* (Hammondsworth, Middlesex: Penguin Books Ltd., 1978).

Sampson, 1965. Anthony Sampson, *Anatomy of Britain Today* (NY: Harper & Rowe, 1965).

Simmons, 1979. I. G Simmons, *Biogeography: Natural and Cultural* (North Scituate, MA: Duxbury Press, 1979).

The Ordnance Survey, 1982. *The Ordnance Survey Atlas of Great Britain* (Southampton: Ordnance Survey, 1982).

B. ON *KAN/MIN* DIVISION

1. English-Language Sources

Beasley, 1972. W. G. Beasley, *The Meiji Restoration* (Stanford: SUP, 1972).

Brown, 1997. Philip Brown, "State, Cultivator, Land: Determination of Land Tenures in Early Modern Japan Reconsidered," *Journal of Asian Studies* 56–2 (May 1997): 421–44.

Chambliss, 1965. William Jones Chambliss, *Chiaraijima Village* (Tucson: The University of Arizona Press, 1965).

Harootunian, 1960. Harry D. Harootunian, "The Progresss of Japan and the Samurai Class, 1868–1882," *Pacific Historical Review* 28 (1959): 255–66.

Ike, 1947. Ike Nobutaka, "Taxation and Landownership in the Westernization of Japan," *Journal of Economic History* 7 (Nov. 1947): 160–82.

Iwata, 1964. Masakazu Iwata, *Ōkubo Toshimichi, The Bismarck of Japan* (Berkeley: UCP, 1964).

Kelly, 1985. William Kelly, *Deference and Defiance in Nineteenth-Century Japan* (Princeton: PUP, 1985).

Mayet, 1893. Paul Mayet, *Agricultural Insurance* (London: Swan Sonnenschein & Co., 1893).

Nakamura, 1965. James I. Nakamura, "Growth of Japanese Agriculture, 1875–1920," in William W. Lockwood, ed., *The State and Economic Enterprise in Japan* (Princeton: PUP, 1965): 262–3.

Nakamura, 1966. ______, *Agricultural Production and the Economic Development of Japan, 1873–1922* (Princeton: PUP, 1966).

Niwa, 1966. Kunio Niwa, "The Reform of the Land Tax and the Government Programme for the Encouragement of Industry," *The Developing Economies* IV-4 (Dec. 1966): 447–71.

Rein, 1889. Johann Justin Rein, *The Industries of Japan* (NY: A. C. Armstrong and Son, 1889).

Smethurst, 1986. Richard J. Smethurst, *Agricultural Development and Tenancy Disputes in Japan, 1870–1940* (Princeton: PUP, 1986).

Shimazaki, 1987. Shimazaki Tōson (William E. Naff, tr.) *Before the Dawn* (*Yoakemae*) (Honolulu: UHP, 1987).

Smith, 1955. Thomas C. Smith, *Political Change and Industrial Development in Japan, Government Enterprise, 1868–1880* (Stanford: SUP, 1955).

Steele, 1989. William Steele, "From Custom to Right: The Politicization of the Village in Early Meiji Japan," in *Modern Asian Studies* 23–4 (Oct. 1989): 729–48.

Steiner, 1968. Kurt Steiner, "Popular Political Participation and Political Development in Japan: The Rural Level," in Robert E. Ward, *Political Development in Modern Japan* (Princeton: PUP, 1968): 221–36.

Totman, 1989. Conrad Totman, *The Green Archipelago, Forestry in Pre-Industral Japan* (Berkeley: UCP, 1989).

Umegaki, 1988. Michio Umegaki, *After the Restoration, The Beginning of Japan's Modern State* (NY: New York University, 1988).

Wigen, 1995. Kären Wigen, *The Making of a Japanese Periphery,*1750–1920 (Berkeley: UCP, 1995).

Wigen Lewis, 1985. Kären Wigen (Lewis), "Common Losses: Transformations of Commonland and Peasant Livelihood in Tokugawa Japan, 1603–1868" (MA thesis in Geography; Berkeley: University of California, 1985).

Yamamura, 1986. Kozo Yamamura, "The Meiji Land Tax Reform and Its Effects," in Marius Jansen and G. Rozman, eds., *Japan in Transition From Tokugawa to Meiji* (Princeton: PUP, 1986): 382–99.

2. Japanese-Language Sources

Aoki, 1968. Aoki Keiichirō, *(Shiryō) Kiso goryōrin jiken kōshōroku* [A record of negotiations in the Kiso imperial forest incident (documents)] (Tokyo: Shinseisha, 1968).

Arimoto, 1968. Arimoto Masao, *Chiso kaisei to nōmin tōsō* [Land-tax reform and peasant unrest] (Tokyo: Shinseisha, 1968).

Chiso kaisei, 1952–1957. Chiso kaisei shiryō kankōkai (Wagatsuma Sakae), comp., *Meiji shonen chiso kaisei kiso shiryō* [Basic documents of the early Meiji land-tax reform] (3 vols) (Tokyo: Yūhikaku, 1952–57).

Fukushima, 1962. Fukushima Masao, *Chiso kaisei no kenkyū* [A study of the land-tax reform] (Tokyo: Yūhikaku, 1962).

Funakoshi, 1954. Funakoshi Shōji, "Chiso kaisei no ringyōshiteki igi" [The significance to forest history of land-tax reform], *Ringyō keizai* 73 (7–11) (Nov. 1954): 1–6.

Hōjō, 1977. Hōjō Hiroshi, *Rin'ya iriai no shiteki kenkyū (jō)*. [A historical study of wooded commonlands (part I)] (Tokyo: Ochanomizu shobō, 1977).

Hōjō, 1978. ______, "Kanrin no seiritsu to shoki kanrin seisaku" [The establishment of *kanrin* and early *kanrin* policy], *TRK KK* 52 (March 1978): 144–73.

Hōjō, 1983. ______, *Meiji kokka no rin'ya shoyū to sonraku kōzō: Nagano-ken Kiso kokuyūrin no sonzai keitai* [Forest ownership and village structure in the Meiji state: the character of government woodland in Kiso in Nagano Prefecture] (Tokyo: Ochanomizu shobō, 1983).

Hōjō, 1988 et seq. ______, "Nihon kindai rinseishi josetsu" [An introduction to forest administration in modern Japan] 1–9, *TRK KK* 62ff, 1988–1999, various paginations.

Hōjō, 1998. ______, "Shimazaki Tōson 'Yōakemae' ni okeru Kiso sanrin jiken no haikei to shite no bungakuteki ideorogii no sokumen" [Literary ideologies as the background of the Case of Kiso Forest depicted in Shimazaki Tōson's novel *Yoake-mae* (Before Dawn)], *TRK KK* 32 (March 1998): 37–47.

Hōjō. [also see Rin'ya seido 1977.]

Hōjō & Ōta, 1980. ______ and Ōta Katsuya, comps., *Nihon ringyō-rinsei no shiteki kenkyū* [Historical studies of Japan's forest system] (Tokyo: Tachibana shoin, 1980).

Kasai, 1984. Kasai Kyōetsu, "Kōyūchi kanminyū kubun no seifu hōshin (1)" [Government policy in the *kan/min* division of *kōyūchi*, part 1], *RK* 425 (37–3) (Mar. 1984): 1–12.

Kasai, 1987. ______, "Kokuyūrin no seiritsu – kōyūrin kanminyū kubun o chūshin to shite" [The establishment of national forest: focusing on the *kan/min* division of kōyūrin] *RK*, 461 (40–3) (1987): 2–4.

Machida, 1978. Machida Masami, "Kisodani sanrin no kan-min yū kubun ni tsuite: Kiso goryōrin jiken no kiin" [Regarding the *kan/min* division of woodland in Kiso: origins of the Kiso imperial-forest incident], *TRK KK* 52 (1978): 174–209.

Machida,1982. ______, *Kiso goryōrin jiken* [The incident in the Kiso imperial forest] (Nagano-shi: Ginkawa shobō, 1982).

Nagatsuma, 1987. Nagatsuma Hiroshi, "Meiji shoki kokuyūrin keiei ni kan suru ichi kōsatsu: tokubetsu kaikei seido michibiki o chūshin to shite" [A perspective on early Meiji government forest administration: the path to the special-funding system], *RK* 460 (40–2) (Feb.1987): 1–8.

Narita, 1996. Narita Masami, "Meiji shonen ni okeru kanrin keiei no keisei katei – Mie-ken Watarai goryōrin no keisei zenshi" [The process of establishing *kanrin* administration in early Meiji: a pre-history of the formation of imperial forest in Watarai (later Mie) Prefecture], *TRK KK* 30 (1996): 47–65.

Niwa, 1988. Niwa Kunio, "Chiso kaisei ni okeru yakibata no haaku" [The handling of swidden in the Meiji land-tax reform] *TRK KK*, 62 (1988): 51–76.

Nihon shiseki kyōkai, 1968. Nihon shiseki kyōkai, comp., *Ōkubo Toshimichi monjo* [The documents of Ōkubo Toshimichi] (vol. VI) (*Nihon shiseki kyōkai soshō No. 33*) (Tokyo: Tokyo daigaku shuppankai, 1968; orig. publ. 1928).

Nishio, 1988. Nishio Takashi, *Nihon shinrin gyōseishi no kenkyū: kankyō hozen no genryū* [*A study of the Japanese forest-administration system: the origins of environmental protection*] (Tokyo: Tokyo daigaku shuppankai, 1988).

Ōshima, 1980a. Ōshima Mario, "Nagano-ken Kiso chihō ni okeru Meiji zenki no kanrin kyōkai chosa ni tsuite" [Regarding the early Meiji surveys of *kanrin* boundaries in the Kiso region of Nagano Prefecture], in Hōjō and Ōta, comps., *Nihon ringyō-rinsei no shiteki kenkyū*: 155–76 (orig. publ. in *TRK KK* 51 (1977).

Ōshima, 1980b. ______, "Kiso chihō ni okeru kanrin kyōkai funsō: Nagano-mura Saruzawa kanrin no jirei" [The dispute over *kanrin* boundaries in the Kiso region: the example of the Saruzawa *kanrin* in Nagano village], in Hōjō and Ōta, comps., *Nihon ringyō-rinsei no shiteki kenkyū*: 177–206; (orig. publ. in *TRK KK* 53 (1979).

Ōshima, 1981. ______, "Kiso sanrin kanminyū kubetsu o meguru sho tairitsu" [The resistance to *kan/min* division in Kiso woodlands], *TRK KK* 55 (1981): 217–30.

Ōshima, 1982. ______, "Kisodani kōyūchi goboku chōsa ni tsuite [Regarding official surveys of "the five trees" in communal lands in the Kiso Valley], *TRK KK* 56 (1982): 147–64.

Ringyō hattatsushi, 1960. Ringyō hattatsushi chōsakai, comp., *Nihon ringyō hattatsushi (jō)* [A history of the development of the forest industry in Japan (vol. 1)] (Tokyo: Rin'yachō, 1960).

Rin'ya seido, 1977. Rin'ya seido kenkyūkai (Hōjō Hiroshi), comp., *Kindai rin'ya seido shiryōshū* [Documents on the modern forest system] (Tokyo: Ochanomizu shobō, 1977).

Shimizu, 1986. Shimizu Yūzō, "Kiso goryōrin jiken no ichi kōsatsu" [An inquiry into the incident in the Kiso imperial forest], *Shinano* 38–12 (December 1986): 26–37.

Tateishi, 1976. Tateishi Tomoo, "Nihonkai hokubu kaigan sakyū ni okeru sabō shokusai" [Sand-control planting on the sand dunes of the northern Sea of Japan coast] *TRK KK* 76 (1976): 150–80.

Tateishi, 1980. ______, "Shōnai ni okeru chiso kaisei to kanrin-kokuyūrin no keisei" [The (1870s) land-tax reform and formation of government forest in the Shōnai vicinity], *TRK KK* 54 (1980): 115–35.

Tateishi, 1983. ______, "Shōnai sakyū ni okeru kanrin-kokuyūrin no hikimodoshi to gyōsei soshō" [The return and adjudication of government forest on the sand dunes of Shōnai], *TRK KK* 57 (1983): 113–38.

Watanabe, 1988. Watanabe Kisaku, "Miyazaki-ken sanrin tokubetsu shobun ni tsuite (2)" [Regarding the special disposal of woodland in Miyazaki Prefecture], *RK* 474 (1988): 27–32.

Index